DESIGN
FOR
HUMAN AFFAIRS

C. M. Deasy

A SCHENKMAN PUBLICATION

HALSTED PRESS DIVISION

JOHN WILEY & SONS

New York — London — Sydney — Toronto

Copyright © 1974 by Schenkman Publishing Company
3 Mount Auburn Place, Cambridge, Mass. 02138

Distributed solely by Halsted Press, a Division
of John Wiley & Sons, Inc., New York

Library of Congress Cataloging in Publication Data

Deasy, C M

"A Halsted Press book"

1. ARCHITECTURE AND SOCIETY. 2. Architecture—
PSYCHOLOGICAL ASPECTS. I. Title.
NA2543.S6D38 720'.19 74-5198
ISBN 0-470-20454-0

CONTENTS

A Word Of Explanation

In 1966 and 1967 I had the pleasure of serving as Chairman of the Committee on Research for Architecture of the American Institute of Architects. As a result of the Committee's activities I came in contact with many of the people whose research was raising such serious questions about the conventional ideas and policies of the architectural profession. In retrospect, the middle sixties seems to have been a particularly fruitful period for architectural theorists. The work of anthropologist Edward Hall, sociologist Herbert Gans, and psychologist Robert Sommer, was evidence of a growing interest within the behavioral sciences in the way people's actions, and to some extent their lives, are affected by their physical surroundings. The architectural faculties at several universities were energetically following up the lead of the behavioral sciences by developing new procedures and new concepts leading to an architecture that was explicitly responsive to the social and psychological nature of human beings. The central idea of this movement, if such a random and uncoordinated activity can be called a movement, was that the fundamental purpose of design was not just to create a building but to provide the physical settings where human beings could live together with maximum amount of useful and pleasurable interchange and the minimum amount of friction and stress. The corollary proposal was that this could be accomplished best through a collaboration between behavioral scientists and designers.

Though my own experience was, and still is, in the field of architectural practice rather than research, this central idea and the evidence that supported it was very persuasive. As a consequence, in 1966, the firm of Deasy and Bolling, Architects, undertook the first of a series of projects in which a comprehensive study of social and psychological factors was used as the first and paramount step in the design of new projects for our clients. Our first effort, planning the interiors of a new office building, seems very simple and somewhat naive today but we learned a lot. Eight years later we are working with greatly improved techniques and a vastly better understanding —and are still learning a lot. During these years we have established, at least to our own satisfaction, that the concept of architects and behavioral scientists working together as a team is not only perfectly feasible but is also capable of producing buildings that are highly responsive to social and psychological needs.

The book that follows is by no means a parade of our projects. It deals in a greatly simplified way with some fundamental concepts of the behavioral sciences that relate to the physical environment and reports some of the research that demonstrates how our actions are influenced by the arrangement of our surroundings. It discusses some of the conventional concepts of the design and planning professions that tend to create behavior problems rather than to eliminate them and outlines the procedures we use to produce what we call behavior-based design. The examples that I use are the ones I know best—our own. While these processes and these examples may lack the polish that might be expected in a more scholarly work, in the complicated world of construction budgets, and building codes they have one great virtue—they work. Since this book is intended for anyone who uses buildings as well as those who design and build them that seems a useful virtue.

While I believe that the approach to architecture that is espoused in these pages is rational and coherent, that doesn't mean that the manuscript describing this approach is thus automatically rational and coherent itself. Far from it. A number of people have contributed to making the text, if not literate at least intelligible, and I would like to acknowledge my indebtedness to them. The top of the list must be reserved for Dr. Thomas Lasswell, Professor of Sociology at the University of Southern California. In addition to being our consultant in social-psychology for the past eight years as well as mentor and friend, Dr. Lasswell has also served as technical editor for this volume. Sheila Segal has served in a parallel capacity as editor in charge of both syntax and good sense. The hand that wielded a very firm blue pencil was equally adept at writing a warm note of encouragement.

Beyond this point the list smacks strongly of nepotism. In reality, it is an impressive display of familial generosity. Two daughters with previous editorial experience, Diana Creighton and Carol Brown, a son-in-low, David Brown, my wife Lucille, plowed through the text and prepared detailed and highly useful critiques, an unrewarded labor of love. The confusion that resulted when all these comments and corrections were transcribed to one manuscript would have defied a professional cryptographer so it is something of a miracle that Marilyn Olson, my very able secretary, ever managed to type a clear and accurate copy.

There are other people who have not been directly involved in the production of this book but have nevertheless earned a special vote of gratitude. The Educational Facilities Laboratory, an affiliate of the Ford Foundation, made grants available to two of our clients that made it possible for us to demonstrate the effectiveness of behavior-based design in large scale public projects. Alan Green, the Secretary of EFL has been particularly helpful to us but I must also acknowledge the support of Ruth Weinstock and Dr. Harold Gores, the President. Their help came at a crucial point in the evolution of our procedures. Lastly, of course, I want to express my appreciation for the interest, support—and the endless patience—of my partners in architecture, Robert Bolling and Donald Gill.

It is my hope that this book will encourage my colleagues in the architectural profession to try this behavioral approach to design. It is, also, my hope that this book will encourage the people who use buildings to expect that their architects will be responsive to their social and psychological needs. If both these ends were achieved we might, just might, make life a little less trying—and a lot more fun.

C.M. Deasy, FAIA
Los Angeles, California
January, 1974

Chapter I

A Strange Way
To Look At Architecture

The idea that the design of the places where we live and work and play has anything to do with the social and psychological stress we experience in pursuing our personal affairs seems curious at first. Preoccupied with the complicated tasks of living, few people give much thought to the way their lives are circumscribed by their physical surroundings. Most of our problems as well as our pleasures stem from the actions and attitudes of other people and most of our attention is focused on them. If we think of buildings at all, it is usually when we are jolted by some outrage such as a stalled elevator or a leaky roof. For the most part buildings fade into an indistinct background, a backdrop for the absorbing activities of human kind.

Buildings and the spaces between them, however, form our cities and towns; we are born in them, live, learn, and work in them, and may ultimately be entombed in them. Viewed as individual artifacts these structures may not seem important but when they are considered together as the artificial habitat our species has created for itself they take on a new significance. Any habitat limits the range, the actions, and the options of the resident species. Like a stage setting that controls the movements of the actors and defines where and under what circumstances they can meet, our habitat channels our movements and influences our contacts.

The influence of our surroundings on our feelings as well as our actions is not the result of impalpable forces that radiate from our environment. In the course of our daily activities we meet other people in a variety of physical settings. Some of these settings are

1

conducive to easy and agreeable contacts while others lead to tensions we could well do without. When someone breaks into the checkout line at the supermarket ahead of you, the natural focus of your annoyance is "pushy people." If the line had been arranged to make line-jumping impossible, however, you would have been spared an unnecessary stress. The chances of redesigning a checkout line to make line-jumping impossible are much better than the chance of reforming the pushy people of the world.

As this example illustrates, while our troubles may stem from the actions of other people, these actions in turn may be prescribed by the arrangement of the physical environment. This ricochet effect is not limited to the annoyances of territorial transgressions. Our habitat also has a lot to do with the way we work with other people. While the accomplishments we achieve as a group are largely the result of the qualities of individual members and the ability of our leaders, group work is more effective in some surroundings than in others.

The endless meetings that seem to be an indispensable prop for the functioning of our society are sometimes more tedious, aimless, and frustrating than they need to be simply because they are not arranged in an effective configuration. Whether it is deciding on a corporate budget, a club outing, or a Little League schedule, a group will be most effective when it is arranged in a formation that is roughly circular, where everyone can see and be seen by everyone else. The importance of this arrangement is easy to grasp when one pictures what would happen to the same group if it were assembled in a narrow corridor where it could only line up in single file. Communication within the group would be difficult if not impossible. Within a short time there would be not one group but several, each settling the problem in its own terms. The villain in this case is not the people—who haven't changed—but the physical setting that forced them into an unworkable social arrangement. The example is rather bizarre but it does prove a point. No leader in his or her right mind would get trapped in such preposterous circumstances. His or her "mistakes" would more likely be of a subtle, perhaps unconscious, kind such as providing too many chairs too loosely arranged. Such a pattern would make it possible for group members who don't really want to participate or to assume responsibility to sit on the outskirts of the meeting, just close enough to drop an occasional monkey wrench into the proceedings.

It is not necessary to construct hypothetical examples to demonstrate how different physical arrangements alter our behavior. One neat demonstration was provided by Stanley Milgram and Hans Toch in the *Handbook of Social Psychology*. Their study was focused on crowd behavior in the lobby of the Brattle Theater in

Cambridge, Massachusetts. After patrons purchased their tickets they joined a queue in a narrow alcove just off the theater entrance. As the line grew it extended down one side of the alcove and doubled back to the starting point in a U. The social priorities of this arrangement are perfectly clear; according to unwritten but immutable rules of decency and fair play, the people in the line were entitled to enter the theater and find seats in the order of their arrival. Late arrivals, obviously, were obligated to follow the line around the U.

The alcove in the lobby of the Brattle Theater was exceptionally narrow, however, and this configuration led to a continuous problem. Our social formations tend to be somewhat loose, under the best of circumstances, so that when the crowd at the Brattle was large, both legs of the U were in contact. Under these circumstances, when the doors were opened and the head of the line moved into the theater they bumped and jostled the end of the line. This contact seemed to be all that was required to release the late arrivals from any obligation they felt to wait their turn. They promptly did an about face and marched in with the head of the line. The people in the middle of the line had to scramble for whatever seats were left. It seems safe to assume that as the middle group entered the theater their hearts were filled with slanderous thoughts of their fellowmen. All this stress and turmoil disappeared, however, when the alcove was widened. The line dutifully pursued its course around the U and harmony reigned.

In his book *Defensible Space,* Oscar Newman deals with another type of behavior that results from the details of building design. Newman, an architect, and director of the Institute of Planning and Housing at New York University directed a three year study to find out how the design of housing projects influences the level of criminal activity within the project.

While most of us probably think in terms of locks, alarms, and barricades as the only feasible means of protection against the threat of burglaries or muggings in the apartment house elevators, Newman's study suggests an entirely different line of defense. In comparing projects which had a high incidence of criminal activity with those which had a low incidence, Newman found that the key difference appeared to be the territorial behavior of the tenants. Where a small number of families shared a common entrance and the units were so designed that it was possible for them to see what was happening in "their" mutually shared public space, a sense of territoriality developed that proved to be a surprisingly effective defense against criminal activity. Territorially-minded tenants reacted to the presence of suspicious strangers and felt no reluctance to call the police if some threat seemed to be developing.

In the large projects which Newman studied, as the extensive and impersonal corridors, lobbies, and stair shafts didn't "belong" to anyone, no one felt responsible for what took place in them. Under these circumstances the public spaces could easily fall under the control of criminal elements. Reading Newman's account of what life is like in low-income high-rise apartments where this situation has developed is enough to make the average reader's blood run cold. What is particularly pathetic is that such situations do not have to exist. As he notes in the comparison of two projects across the street from each other, the difference is solely the result of the way they are designed.

> *The one hundred and fifty New York families trapped in apartments that open onto the double-loaded corridors of a seventeen story high-rise building—whose elevators, fire stairs, hallways, and roofs are freely roamed and ruled by criminals—find it hard to believe that the project across the street, composed of three to six-story buildings in which two to three families share a hallway and six to twelve an entrance, actually accommodates people at the same density and could be built at the same cost. The families in the seventeen-story building find it incomprehensible that both projects house families at equal densities and that the design differences between the two projects are predominantly the result of the whim of each designer. ...It seems unforgiveable that high-rise projects would have been designed to make their inhabitants so vulnerable, when projects across the street were able to avoid these problems simply by not creating them in the first place.*

Newman's study seems to afford a flawless example of how behavior differs in different settings. To be sure, it is the people, the criminals, who enact the behavior, not the buildings, but one arrangement makes criminal activity easy while the other makes it difficult.

There is one aspect of Newman's statement that is particularly telling for anyone in the field of planning and design—the charge that the designers of the high-rise project selected a seventeen-story solution as a personal whim. "Whim" may be too derogatory a term to apply to that decision, but it does seem clear that the designers could not have had the faintest idea of the disastrous effect their decision would have on the tenants. Regardless of the other talents they brought to their architectural assignment, they obviously did not bring an understanding of environmental influences on human behavior.

Our judgment may sound like a very harsh indictment of these designers, but the astonishing fact is that the same thing can be said about the great majority of all design projects, public and private, large and small. While the nature of the surroundings in which we live clearly affects our actions, our contacts, our sense of identity, and our effectiveness as human beings, the traditional processes of the design and planning professions do not deal with social factors in a rational and systematic way. Architects and planners use a number of measures for evaluating buildings and plans, such as function, stability, cost, and aesthetics. Recently they have also begun to judge them on the basis of their ecological impact. The idea that buildings and communities should be judged first of all as behavior settings is just beginning to be accepted. For many people this will seem to be a strange way to look at architecture.

"We Shape Our Buildings, Etc."

In designing a home there are a number of functional relationships that a housewife will want to work out very precisely. The relationship between the place where the car is parked and the kitchen storage cabinets will have a great deal to do with how far heavy parcels will have to be carried. In the same fashion, the distance between the place where the food is prepared and the place where it is eaten controls the number of miles the family will log annually in trotting back and forth. This kind of analysis can be applied to a host of other relationships in order to produce a convenient and functional plan. In designing a house there will also be a complicated set of decisions to make about technical components such as the structure and the mechanical systems. These must be safe, reliable, and desirable behavior and decrease the possibility of undesirable behavior, the pattern of the materials.

Having made all these choices, the house should be convenient, functional, safe, easy to maintain, and beautiful. All of these things may seem to be enough to ask for a house, but in reality the apparently comprehensive traditional formula sidesteps the most important point of all. The basic purpose of a house is to provide a place where a group of individuals can co-exist with a minimum of friction and a maximum of satisfaction. While there is no plan that will guarantee perpetual domestic tranquility, there are some arrangements that will make family life distinctly easier and much more enjoyable than their alternatives. Designing a house with this in mind, we would start by defining the social interrelationships to be dealt with rather than the operating functions. If we can arrange to increase the possibility of desirable behavior and decrease the possibility of undesirable behavior, the house would be conceived as a behavior setting.

While the systematic use of social psychology is relatively new in the design field, it has been urged for some time by a number of leaders in the behavioral sciences. The most practical as well as the most lucid comment on the way that architectural arrangements can be used to influence behavior doesn't come from a professional social psychologist or a cultural anthropologist, however, but from a master politician, Winston Churchill. His observation, "We shape our buildings and afterwards our buildings shape us," has been endlessly repeated in the literature of architecture and in the oratory of architectural gatherings. Taken by itself it seems to imply that the design of buildings is a noble endeavor that will shape a better life for multitudes unseen and unknown, an idea that has great appeal to architects. Regrettably, Churchill had no such lofty message in mind. Speaking in 1943, after German bombers had destroyed the House of Commons, he was urging its immediate reconstruction and his speech demonstrated a crystal-clear understanding of how the shape of a meeting chamber influenced party discipline:

> *The semi-circular assembly, which appeals to political theorists, enables every individual or every group to move around the centre, adopting various shades of pink according as the weather changes. I am a convinced supporter of the party system in preference to the group system. The party system is much favored by the oblong form of chamber. It is easy for an individual to move through these insensible gradations from Left to Right, but the act of crossing the floor is one which requires serious consideration.*

Having switched parties—"crossed the floor"—twice in his career, Churchill undoubtedly had a clear idea of the restraint that this configuration imposed on impulsive switching. While the virtues of the party system as a means of establishing national policies may be open to some question, there is no doubt that if one wanted to strengthen that system and to clarify the distinction between the "Ins" and the "Outs", a face-to-face arrangement would help. This is the pattern of confrontation rather than of cooperation.

Churchill's insights didn't stop at this point. He also showed an acute awareness of the psychological effects of crowding.

> *The second characteristic is that it should not be big enough to contain all its members without crowding...If it is big enough to seat everybody, 9/10ths of all debates will be conducted in the depressing atmosphere of an almost empty Chamber. The essence of good House of Commons speaking is the conversational style, the facility for quick, informal interruption and interchange. But the conversational style requires a fairly small space, and there should be on great occasions a sense of crowd and urgency...There should be a sense of the importance of much that is said, and a sense that great matters are being decided, there and then, by the House.*

Churchill's whole speech demonstrates a surprising grasp of an aspect of design that has only rarely been considered in a systematic way. To be sure, an architect would have a host of other matters to consider in designing a new House of Commons: lighting, acoustics, and circulation as well as the materials and technology of construction. He could, however, solve all these problems admirably and still miss the main point entirely if he failed to define his objectives in terms of human behavior. Unfortunately, such failures are commonplace. The design professions have no systematic process for defining behavior goals and no tradition of research that would supply the necessary information.

Charles Abrams, a planner of great perception and wide experience has commented on this astonishing void in the design process. After interviewing the architect for the Cleveland Zoo, he concluded that no human animal ever received as much concerned attention as the zoo animals. Before a line was drawn on the plans, specialists from all over the world were consulted, not only on feeding and climatic tolerances, but on such crucial behavior factors as social organization, dominance patterns, territoriality, and mating habits. The logic of this course is evident. Regardless of how well the zoo might perform in other ways, if it failed to provide an environment in which the animals could not only survive but could function in their natural way, it would have to be judged a total failure. While it may be an unpalatable analogy, it is hard to understand why the needs of human beings, who are at least as sensitive and complicated as banded armadillos and lesser auks, would not receive an equally comprehensive study in the design of the human environment.

This indictment immediately raises a perplexing question. If the manufactured surroundings we live in are so inadequate, why isn't it immediately apparent to everyone? The answer is found at several levels. Our surroundings are seldom so obviously restrictive that they incite revolt, though such a result is not completely unknown. Their effect is a matter of degree. While we can clearly perform better in one environment than another, it would not be apparent unless we had some way of comparing the two, an opportunity that we rarely have.

We are a supremely adaptable species and we can tolerate a considerable amount of discomfort and inconvenience, particularly if bringing about a change would appear to require more effort than adapting. As Rene Dubos, the distinguished micro-biologist, has pointed out, man's capability in this regard is so extraordinary that it is entirely conceivable he could adapt to conditions that would destroy the values that make him unique. While Dr. Dubos' statement is intended to be a shocker, the idea has some intriguing aspects. It suggests the dazzling possibility that mankind could be induced to adapt away from some characteristics that he would be better off without. Unfortunately, no one can state with authority what an improved model should be like or how to bring it into being. For our purposes, a knowledge of human adaptability should be used to provide the maximum opportunity for us to be at our best and the fewest reasons for us to be at our obnoxious worst.

The last reason we fail to recognize pressures from our surroundings follows from our lifelong training in accepting the restraints of social conventions. Part of the price we pay for living in an organized society is the acceptance of certain rules. Some of these are necessary to reduce the stress of living closely together while others are rather pointless holdovers from other times. In either case, we are not likely to know the reasons for the rules. It is the nature of such conventions that they are simply accepted. Our convention that we pass oncoming traffic to the right rather than the left is such a crucial one that it has been embodied in law. Its usefulness is beyond dispute. Any attempt to apply a laissez-faire philosophy to freeway traffic, with everyone free to follow his own dictates, would produce chaos beyond imagination. Other conventions are not nearly so useful. Our convention that a single family home must be separated from other structures on all sides produces a lot of useless yard space that is expensive to buy and exasperating to maintain, but we accept this pointless burden as a necessary, but unexplained, evil. When we consider the conventions of our cities, there is a tendency to accept them, flaws and all, as the forms our society has evolved to make social living possible.

Paul Ylivsaker's statement, "We see God in Nature, Man in Cities," tends to support the view that our cities are a reflection of our society, but it is, in one sense, misleading. What we see reflected in our cities is not Man, but *some* men. The cast of characters who shape cities is a complicated one, to be sure, but the assumption that the nature of the buildings we use or the communities in which we live faithfully reflects the needs or the nature of all of the inhabitants is sadly inaccurate. We adapt to them as conventions which we must accept.

It is the current mood of our nation and of our time to re-examine critically many forms and systems which we have long taken for granted. In that process it is inevitable that the nature of our man-made environment and the performance of the people who shape it will be subjected to a searching analysis.

Putting Down the Architect

The inadequacies of the built environment in social terms have been, as might be expected, a special concern of the sociologists. Indeed, it has become almost a hallmark of the literature on society and environment that the author detour from his principal objective long enough to direct a withering fire into the camp of the enemy, the planning professions. In articles and interviews, his job is not finished until he has lobbed a few mortar rounds in the general direction of architects and planners.

To describe the attack of sociologists on the planning professions as warfare would be inaccurate since, at the moment, all the shooting is coming from one side. Designers and planners do not seem to be shooting back and it may be that they are not aware that they are under attack. If this is the case it is regrettable. Although some of these assaults only demonstrate that behavioral scientists are as capable of being obtuse as anyone else, their main claim—that the planning professions have subordinated vital human values to somewhat arbitrary professional values—is well documented. The professions would serve the public better by giving these charges very careful consideration. Pointed barbs like the following are coming thick and fast.

> *...most architects don't have the foggiest notion how society works, how people live, and how they want to live. (Herbert Gans, Professor of Sociology at Columbia University and author of "The Levittowners," "The Urban Villagers," etc.)*
>
> *They have encouraged the development of an extensive self-congratulatory system within the design professions. The present system is reasonable if architects are giving themselves awards for sculpture but not if the awards are intended for buildings in which certain activities will take place. (Robert Sommer, Chairman, Department of Psychology, University of California, Davis and author of "Personal Space," "Design Awareness," etc.)*
>
> *Part of the difficulty stems from the fact that the builders, if not planners, ignore the critical reality that cities are primarily social organizations—that they are only secondarily collections of concrete, steel, and wooden structures. (James Birren, Director of the Gerontology Center and Professor of Psychology at the University of Southern California.)*

Robert Gutman, Professor of Sociology at Rutgers University, is more specific in his comments. Since he has studied the sub-culture of architects and planners at length, his observations on the disparity between their idealized solutions and the realities of human use carry more weight than most:

> *...the history of contemporary buildings is full of examples of areas set aside for parks which have never been used for this purpose. The children who were supposed to play in the grassy space can be found tumbling in the adjoining sandlot; instead of sunning themselves on the benches, the men are gossiping and drinking at the saloon around the corner.*

While these assaults are painful for an architect to absorb quietly, they are much less persuasive than the volumes of research data that, in a quieter way, tell the same story. For the most part this material—hidden in obscure journals—does not deal directly with architecture or planning at all. It simply relates the results of painstaking investigations of the way people interact in certain situations or the kind of preferences they express concerning their surroundings. The importance of this information only becomes clear when it is compared to the conventional practices and assumptions of design professionals and their clients.

A number of observers, for example, have commented on the human preference for facing a conversational partner, not eyeball to eyeball, but at an open angle. It is the most effective way to employ all the devices of communication—gesture, expression, and voice, without the strain of face-to-face confrontation. Yet the public seating in this country, in parks, bus stations, or airports, rarely reflects this fact. For the most part it is arranged in formal lines that seem to assume that strangers never speak. Under such circumstances they seldom do.

A research team at Penn State undertook the ambitious project of rotating a group of families into the same dwelling for a period of weeks in order to study their living habits and patterns of use in the same surroundings. Their report is a fascinating document for anyone who is interested in human beings. For the serious designer, who feels some commitment to produce better housing for the American family, it offers a host of clues on how this might be done. They are clues only, however, not recipes. The designer must infer solutions from the described behavior. A simple example can be constructed from the behavior of young fathers who showed a tendency to stay away from the house when they had office work to do unless they could obtain a location for privacy when they were at home. If we accept the premise that it is desirable to have a father around the house, then it seems reasonable to propose that the house be arranged to offer him some place where he can be alone at times. This would be a modest concession, but it is a feature that is notably lacking in the typical American home.

Dr. Richard Myrick and his associates developed some equally useful information on an entirely different topic in a study of Washington, D.C., high schools. Their attention was focused on the movement patterns and social habits of students while they were outside the classroom. This is a matter that has received very little attention from planners, who have tended to focus on the formal educational experience itself. Time spent outside the classroom is often regarded as time wasted. The Myrick study indicates that this neglect of activities between classes is an unfortunate oversight. Not only are the needs of the students, who are the primary users of school buildings, overlooked but the time between classes also proved to be an integral part of the education process. Those gaggles of gossiping students that clog hallways between classes spend a part of their time discussing classwork and future assignments. Dr. Myrick also points out that while cross-fertilization between different academic disciplines is an often stated goal of education, the actual layout of school plants, the rigid departmental segregation, tends to defeat this goal. On the basis of the findings of this study, anyone with an intimate knowledge of school planning practices in this country would have to conclude that much of the attention of architects and school administrators alike has been misdirected.

While serious studies that question the competence of the planning professions are disturbing, what hurts the most are challenges to the designer's assumption that design quality—the visual characteristics of surface, pattern, and form—is a very important factor in the human habitat, an abstract "amenity" that can't be measured but which contributes to our enjoyment and sense of well-being. The team of Foote, Jughod, Foley, and Winnick (*Housing Choices and Housing Constraints*) has undermined this assumption in a study of why people like or dislike the places in which they live. Physical amenities seem to have little to do with their preferences.

> *When the housing consumer evaluates his neighborhood satisfaction his central concern is neither geographic site nor physical characteristics. Among consumers satisfied with their neighborhoods—the chief reason for satisfaction seems to be the social characteristics of their neighbors. Among consumers dissatisfied with their neighborhoods—the basic cause seems to be again the social characteristics of their neighbors.*

This doesn't quite tell the whole story, for there are situations—particularly when we encounter a new environment, buy a new house, or pick a new apartment—where design characteristics influence us greatly. What the study does emphasize is the relative unimportance of design characteristics when compared to our personal preoccupations when we are alone and our focus on social responses when we are with others. This is hardly a surprising revelation. It is essentially the message of the famous Hawthorne study of 1927, but in the forty-five years that have intervened, the implications of this fact have not been absorbed into the dogma of the people who design our physical environment.

A last thrust comes from Festinger, Schachter and Back (*Social Pressures in Informal Groups*).

> *The architect who builds a house or designs a site plan, who decides where the roads will and will not go, and who decides which directions the houses will face and how close together they will be, also is, to a large extent, deciding the pattern of social life among the people who will live in those houses.*

Festinger, et al., may be claiming too much. It is hard to believe that socially competent adults with the mobility and communications that are available today would permit themselves to be constrained to any great extent by some planner's decisions, made without conscious knowledge of their social effects. People may be hampered by such decisions but not ultimately controlled by them. For the elderly, the ill, and the infirm, for little children, and for the newcomer, it is a different story. They can be as effectively isolated by the thoughtless acts or omissions of the planner as if they were sealed in a tomb.

The charge that architects have some responsibility for the social content of the lives of the people who use their buildings, not only in subdivisions and housing tracts but wherever they come together, is a serious one. It is a responsibility that most architects probably would be reluctant to assume. Things are complicated enough as they are. The point made by behavioral scientists who have been most concerned with this subject is that, like it or not, it is a responsibility they have. Robert Wehrli, architect-psychologist at the University of Utah, has summed it up more bluntly than most:

> *...for whenever he builds he controls or guides human behavior. When enlightened as to the effects of the physical environment upon behavior, he designs by intent; but when ignorant of these effects, he designs by default.*

an argument leading to the curious conclusion that planning, where a large share of the crucial decisions are made by default, must really be characterized as non-planning.

In assembling the data for this book I have accumulated a substantial stack of file cards which, like the preceding examples, deal with the general wrong-headedness of architects and planners. All of them originate with sociologists, psychologists or anthropologists. None of the cards, however, contain any hint as to why these design professionals perform as they do or what action these eminent social scientists would suggest in order to bring about a constructive change. That is a disappointing omission. The social sciences have all the skills and techniques necessary for finding out *why* designers act as they do. If their practitioners would use these techniques to study the design field and the special constraints under which the design disciplines work, they would not only explain some puzzling phenomena, they might also set the stage for a change in direction.

It is not my purpose to offer a defense for the fields of planning and design or to shift their blame to someone else. Any sins of commission or omission that they may have committed are highly visible, immortalized in steel and concrete. Regardless of their expressed concern and their stated commitment to serve the community, the evidence indicates that they have been sadly remiss in developing design procedures that respond to the human needs described by the behavioral scientists. There is not much hope, however, that fretting about it will do any good. New processes will have to be defined, and these processes must recognize the designer's peculiar problems if they are to have any effect.

The designer's primary problem stems from a unique role relationship with the people he is supposed to serve. Like other professionals, he is traditionally, financially, emotionally, and legally bound to a client. However, a paradox arises from the fact that in many instances the client who employs him is not representative of the people who will use his commissioned creation, nor is he even a reasonable surrogate for the users. While it may be true that the designer sometimes fails to serve the needs of his informal client, the users, it is usually not true that he fails to serve the expressed needs of his formal client. If he failed in that respect, he would probably starve to death. As a consequence, even though he may be keenly aware of user needs and may have grave misgivings about the policies of his formal client, there is little hope that he can really serve both of them well unless he has some universal technique for defining user needs and relating them to the overall goals of his formal client. Nothing in the designer's conventional arsenal of techniques is capable of resolving that knotty equation. That doesn't imply that a solution that is relevant for both groups is impossible. In fact, the major theme of this book is to describe a mechanism for elevating the needs of users ordinarily unseen and unknown by the designer to the level of mandatory design criteria. In the special language of the social sciences we could say that the designer must assume a new role with the user as a reciprocal role partner. In more prosaic terms we could describe this mechanism as a process that forces designer and client alike to recognize that buildings, towns and cities are inhabited by real, live human beings whose needs and concerns are frequently ignored.

New Rules for an Old Game

In defining a more precise set of human goals for the planning fraternity and describing a process that would make these goals attainable, it is necessary to fix some point of departure. In spite of the criticism that they have received, these design disciplines display impressive talents in terms of technical competence, organizational skills, creativity, and the ability to synthesize complex sets of data. The phase of their practice that requires reform is the area of pre-planning, gathering information about people and establishing human criteria that will determine the nature of the design and define the purpose of the product. To be useful, the procedures used to establish these criteria must be realistic, not theoretical, and must have direct application to the realities of the construction field.

In order to give purpose and direction to any search for information it must have an underlying theory; it must be based on some underlying concept or philosophy. The concept I would suggest is that the only worthy basis—indeed the only rational basis—for designing anything would be to make the human beings more effective. On the surface this may not seem like much of a change, since almost everything with which architects and planners have traditionally been concerned can be stuffed under that generous umbrella. New street patterns that make movement and communications easier contribute to human effectiveness. Better lighting, better temperature control, and the efficient arrangement of work stations all contribute to our effectiveness and have been a major concern of the planning professions.

There is one crucial factor, however, that has long been ignored in the traditional design approach, as our friends the behavioral scientists delight in pointing out. That is the fact that the social and psychological milieu in which we live and work has far more to do with our effectiveness as human beings than our physical environment. Productivity and creativeness are not unique to new buildings, regardless of how well equipped and well designed they may be. In what is something of an embarrassment to designers, these elusive characteristics seem to blossom quite as well in derelict stores and abandoned car barns as anywhere else. The best teaching does not always occur in the model classroom, the best cooking does not necessarily come out of the most stylish kitchen, and healing miracles are not confined to the newest hospitals.

It is apparent that the designer who is seriously concerned with the question of human effectiveness must widen the scope of his interests to include a complex new set of human factors. If personal motivation and group interaction are elements that affect our competence as human beings, the designer has a clear responsibility to create environments that will do as much as an environment can do to recognize and accommodate these factors.

There is some danger that the phrase "making people more effective" might be misconstrued to mean merely making people more efficient or productive in some quantitative sense. While this certainly is one possible result, the intention is much broader. The phrase is intended to cover the whole range of human activities that might be influenced by the designed environment, and to support the individual's efforts to achieve whatever goals he has set for himself, or at least to hinder them as little as possible. In her book *With Man in Mind,* Constance Perin deals with this same concept in terms of "competence." "The concept environmental design might organize its data around, as it measures and estimates the consequences of what it does and proposes to do, is that of the *sense of competence* people have in carrying out their everyday behavior—visiting, working, playing, learning, shopping, meditating, cooking." Whether it is competence or effectiveness, the idea is the same. Both define a new focus for design.

In marshalling information about the human species as a preface to the actual task of planning, one useful convention is to assume that man has several sets of dimensions: anatomical, physiological and behavioral. This convention is particularly useful because it fits traditional patterns of thought.

Architects have been dealing with the anatomical dimensions of humankind since the days of ancient Sumer and have an accurate fix on them. While people vary considerably, the range is predictable and architects know that designing a doorway only four feet high would lose them a lot of friends. It's true that they may not always use this information judiciously. There seems to be a deplorable amount of wishful thinking about the dimensions of the human rump, leading to a lot of unnecessarily uncomfortable seating. Failure to use available information and just plain mistakes can also occur in less obvious ways. The first time I saw Dulles Airport (the design of the late and greatly gifted architect Eero Saarinen) in Washington, D.C., rough boards had been wired across the widely-spaced bars of the stair rails to prevent some child from plunging to his doom. At the same moment similar boards were wired to the handrails of the penthouse on an office tower we had designed in Los Angeles, so the matter of physical dimensions is one that has been impressed strongly on my mind.

Man's set of physiological dimensions is somewhat more complicated but reasonably well defined. In the physical sense, that is what shelter is all about. The human tolerance for heat, cold, air movement, and the quality of light necessary for good vision have all been the subject of exhaustive study. For the most part environmental designers do a good job with this set of dimensions, though again there are glaring lapses. When you see a modern building with aluminum foil pasted on the windows or cardboard deflectors taped to the air conditioning outlets you can be sure that someone hasn't kept the physiological dimensions of the human users in mind.

There are other, more subtle physiological factors that have received less attention. Our echo-ranging ability, the curious technique by which we locate ourselves in relationship to other things and other people by interpreting the source and distance of sounds, was probably a crucial talent in our jungle days and is no less important in the streets of the city. In the general uproar of city life it is a talent that seldom gets a chance to function. We also have a talent for interpreting tactile sensations that could be used much more than it is. Traffic engineers, through their use of dot markers that warn us when we are drifting out of a freeway lane, have exploited this tactile sense to some degree but it has not been used widely in the building field. There are many ways in which these and other physiological factors could be used to make our environment both safer and more informative if they were widely understood and generally employed. A handbook of physiological dimensions similar to the handbooks of anatomical dimensions that are in common use would be a blessing for both the people who design buildings and the people who use them.

When we come to man's behavioral dimensions, the situation is radically different. With all the mountains of data that have been assembled by the behavioral sciences, there is nothing like an authoritative statement that could be used as a simple guide on "How to Plan for Human Beings." While there are certain concepts that seem to be generally accepted, concepts dealing with such characteristics as security, recognition, membership, and so on, they are meager fare for the designer. Since he must ultimately produce a three-dimensional construction with a precise set of dimensions, generalizations about the nature of human nature are not much help. He needs specific criteria and they are simply not available in handbook form at the present time.

There are some exceptions. Certain psychological concepts, particularly those dealing with perception, are in common use by the architectural profession and have been for centuries. In one sense, the designers of the ancient Greek temples and the architects of the Rennaissance churches must be counted among the first experts on perception. They clearly understood certain perceptual phenomena and freely altered the form of statues high on their buildings so that they would appear normal when seen from the ground. They forced the perspective within their churches to make them look longer or higher than they were and used the principles of contrast to induce people to see what they wanted them to see. These particular practices may not be much employed today, but the knowledge that we perceive objects better when they are seen against a contrasting background, that certain forms are comprehended more readily than others, and that certain kinds of lighting improve our ability to understand the nature of materials, are all still a fundamental part of every architect's design technique.

So far as most other aspects of man's complex nature are concerned, there seems to be no reason to hope that we will soon, or ever, possess a set of behavioral dimensions that approach even remotely the reliability of our tables of anatomical dimensions. It is probable that this fact, as much as anything, accounts for the tendency of planners and designers to neglect these confusing areas of human concern. Dealing with the crisp realities of the physical world offers some positive advantages; the elements of the problem can be defined precisely and the success of the solution can be measured in quantitative terms. In contrast, human factors are subtle, often appear to be contradictory, and can only be measured in broad and general terms.

The fascination with physical solutions is not limited to the design field. Faced with confusing social and personal problems, managers in the field of private enterprise and managers of our public affairs repeatedly turn to physical solutions rather than face the complex problems of managing social change. In many instances such responses are totally inadequate, as the dramatic failure of some grandiose public programs has demonstrated. In terms of our manmade environment they are often particularly inappropriate. To a sometimes unappreciated extent, man carries much of his environment with him, not on his back but in his head. In a wry twist of terms, it is not only appropriate but necessary to think of the manmade environment as *man's mind-made environment.* As a consequence, the physical planner must adopt some systematic procedure for dealing with the behavioral dimensions of interacting humankind.

At the present time the only hope of incorporating authoritative social and psychological information into building programs lies in a partnership between the design professions and the behavioral sciences. That is the most reasonable method that the the design professions have for obtaining access to the enormous body of data that has been accumulated about human behavior. Even more important, it provides access to the data-gathering *methods* that these sciences have developed. This is far more significant than it may appear to be. In view of the fact that the environmental designer normally deals with a specific case in a specific locus it is imperative that he work with the factors that are unique to that case. Only through the special perspectives and the research methods of the human sciences can a with reasonable confidence designer presently hope to discover what those factors are.

The collaborative approach that I am suggesting is not a fantasy. It has been tried and it works, though not without some problems. The implausible hybrid, Behavioral Scientist X Designer, has shown that it is capable of resolving complex human problems in very practical ways and holds the promise of radical improvements in both the planning process and the planning product. In genetic terms, it has demonstrated remarkable hybrid vigor.

A Tentative Sort of Marriage

The idea that architects and planners should collaborate with behavioral scientists for the ultimate benefit of building users and city dwellers is not a new one. The potential advantages of such a union have been discussed for some time. Not much progress has been made in putting this idea into effect, however, probably due to the lack of a well defined process for joint action. Simply putting a representative of each of these disparate disciplines in the same room with the vague hope that something useful will occur is not likely to be productive of detailed solutions to specific problems. It may be an enjoyable and stimulating session, but it will not necessarily be a fruitful one. They would, at the outset, face a formidable language barrier. Like most professions, each of these disciplines has developed a special vocabulary that is capable, on occasion, of making them virtually unintelligible to the rest of the world. Each has a set of attitudes and a value system that is unique to the breed, and each has a set of processes and capabilities which may be utterly mystifying to the other.

In studying the relatively few cases of successful collaboration that are available, there seem to be certain factors that are essential: both parties must have a sincere respect for the capabilities of the other, the designer must feel a genuine commitment to base his design on the data provided by his behavioral colleague, the collaboration must be initiated at the outset of the project, and the liason must be more than temporary. Of these, the designer's commitment is undoubtedly the most important. In those instances where independently conducted behavioral studies were offered to designers to guide their work, the evidence indicated that the finished product reflected little, if any, of this special information.

In its simplest form, collaboration takes place when a designer retains a behavioral scientist as a consultant. If the designer is capable of defining his problem clearly this can be surprisingly effective. The knowledge about human beings that a social psychologist has at his fingertips, and the additional knowledge that is available to him through a data search, can be impressive to anyone who does not have this same intimate understanding of the behavior field. It is such a productive process and so simply undertaken that it is regrettable that it is not used more often. My own initiation into man-environment studies took place in this way. Compared to the modest amount of time and money invested in consultation, the practical benefits that accrued to my clients and to the ultimate users of the building were a bonanza. The initial results were so useful that they led directly to the more sophisticated methods that we have since learned to employ.

Beyond the level of consultation and data search, the behavioral scientist can participate in both project management and information gathering. He can define the kind of information needed by the architect or planner to resolve human problems and he can direct the field work that is entailed in getting that information. He can participate, finally, in evaluating the different responses that the architect can make to such information. In practice this can range from relatively simple surveys of user preferences to ambitious studies leading to the development of social-psychological profiles of the various people or populations affected by a planning decision.

Our own work reflects all these possibilities. Since 1966 we have worked with Dr. Thomas Lasswell, Professor of Sociology at the University of Southern California. His collaboration has ranged from consultation and research to rather complex studies of specific communities. Whenever it has been possible to use it, the most productive process has been one in which the information necessary for a particular project is gathered by systematic observations of similar facilities in use, followed by an interview program with carefully selected samples of the different populations that will be influenced by the project. In one way this procedure seems to depart from other practices of which I am aware. Instead of confining interviews to the normal questions about user preferences, the scope has been broadened to include more fundamental matters such as strains, goals, attitudes, values, images, and motivations. Some writers would challenge the assumption that an architect can make a relevant response to such abstract factors. I concede that they are difficult to deal with, but they also lie at the heart of an architect's effort to make his work useful in something more than an ordinary way. Taken together with the observations on human response to a given environment and the answers to more prosaic questions about where the housewife wants her kitchen, these abstract issues are necessary to complete the description of the problem the designer has to solve.

The contribution of others to the development of the behavioral approach to design has been crucial. Robert Gutman's series of papers identifying the special attitudes and values of the subculture of architects and planners jolted me into the realization that professional groups are capable of developing some rather pecular points of view. "Dorms at Berkeley," a study of student behavior in one of the high-rise dormitories at the University of California made by architects Sim Van der Ryn and Murray Silverstein, was another eye-opener. Using both observations and interviews, this team rubbed my nose in the fact that people may use buildings in ways that are completely different from those that the designer has assumed. Even more important, they demonstrated that their techniques could be used to design better buildings as well as to evaluate existing ones.

The "Windowless Classroom" study undertaken by the Architectural Research Laboratory of the University of Michigan, raised more questions; if it made no difference whether the windows were in or out, then it seemed entirely possible that my colleagues and I had been spending a great deal of time worrying about some aspects of the environment that no one else cared about. To reinforce this unsettling idea, there was a continuing flow of man-environment studies emanating from the Ethel Percy Andrus Gerontology Center at the University of Southern California and the Architectural Psychology program at the University of Utah. In case these sources might not in themselves persuade a practicing professional that it was time to change, there were those clear, ringing messages from such astute scholars and vocal critics as Robert Sommer, Edward Hall, and Herbert Gans to prod him along.

While the action on the behavior-architecture front has largely been confined to academic circles as the above list suggests, there are a few exceptions. Ewing Miller, an architect of Terre Haute, Indiana, has been working with psychologist Lawrence Wheeler for a number of years on a continuing series of projects. Architect Brent Brolin and sociologist John Zeisel have established a joint practice, the ultimate form of collaboration. Their proposals for housing prototypes for the West End of Boston, derived from Herbert Gans' exhaustive sociological study of that area (*The Urban Villagers*) are as good an example as one can find of how effectively an imaginative team can respond to the social needs of a special group.

The practical example that pleases me most, however, is offered by Louis Sauer, a Philadelphia architect, and his colleague, David Marshall. It is an admirable example because it demonstrates how little it sometimes takes to change a plan from something that works to something that works well, and how much a team of architects who are committed to the real needs of the users can accomplish without any counsel or assistance at all. Their project was the design of low-cost housing for a public agency in which the criteria, both in cost and planning objectives, were established by the agency staff. Meeting these criteria was not particularly difficult, but the design team could not escape the uneasy feeling that this professional staff might not exactly reflect the habits and life styles of their low-income clients. On this point they were precisely correct.

Without any elaborate interview processes, keypunch operations, or computer printouts, Sauer and Marshall were able, in a very limited series of meetings with prospective tenants to determine a surprising number of facts about neighborhood patterns and preferences that altered their original concept in fundamental ways. The before and after plans don't seem radically different. The buildings are the same size, of course, reflecting a similar construction budget. The distribution of the space between areas is altered somewhat but the one thing that would be most likely to register on the mind of the casual observer is that the living room that was facing the private yard in the rear is now facing the street. This is not much of a change, on the surface, yet it reflects the fact that social life in this ghetto was not like that in suburbia.

The living-room-to-the-rear arrangement that is so often seen in suburbia provides for a private family life sheltered *from* the street. In the neighborhood where Louis Sauer and David Marshall were working, a living room is the center of an easy social activity, for adults only, based on visits from friends and neighbors and oriented *to* the street. The inward-looking, self-sufficient arrangement of suburbia would have been very inappropriate for these ghetto families.

If such a simple endeavor as the Sauer-Marshall study can prove to be useful in developing an environment that is supportive of human effectiveness, it is painful to consider how many similar opportunities are passed over by the planning professions every day.

While it is true that Sauer and Marshall might have accomplished even more with a more elaborate study, consider how much worse off they would have been with no study at all!

Perhaps this last example serves to illustrate that this book is not committed to any rigid process or precise theory. If it were intended to be a scholarly treatment of the subject it would be *pro forma* to state a general theory and to proceed to support it with persuasive evidence. No such statement is possible, or even desirable at this time. The information at hand seems to indicate that, so far as the planning professions are concerned, the only general theory that makes any sense is that each case is different, and the architect or planner who seriously undertakes to satisfy the human needs of his constituents must deal with each situation as though it were unique, for in many ways it is. In time, enough information may be assembled to make it possible to draw some general conclusions that will serve as reliable guides in the planning process but that time is in the distant future.

The pages that follow are intended as a guide to some of the methods that can be adopted now to improve planning for human effectiveness. It is my hope that they will prove useful to my colleagues in the design and planning fields, to the clients, administrators and managers who make so many of the crucial decisions about our built environment, and for anyone else who has an interest in their personal effectiveness and their relations with other people. These methods may fall somewhat short of meeting the exacting standards of pure science but in the complex and confusing real world of construction budgets, building codes, and time schedules they have one outstanding advantage for the human user—*they work!* They are based on the premise that the planning professions must shift the focus of their concern to human behavior and to design, in Constance Perin's term, with man in mind.

Chapter II

How The Behavioral Scientists
Look At The World

It would, no doubt, be interesting at this point to move on to a discussion of practical applications of behavioral research in architectural planning. The opening chapter made some fairly generous claims in that regard and it would be perfectly normal for some readers, in a "put up or shut up" mood, to expect an immediate demonstration of results. Anyone who feels that way can skip this chapter and move on to the pragmatic discussions that come later. They may, however, find these discussions a little hard to follow. The point of behavior-based design is that human beings have certain attitudes and drives that are largely ignored in our existing urban environments, but it is difficult to evaluate proposals to make these environments better unless there is agreement on what these attitudes and drives are. My suggestion that involving users in the planning process may mean more to them than the quality of the plan has a zany ring to it unless one is aware that the desire to be recognized as an individual is a deep-seated human drive. In the same way, my suggestion that architects should not attempt to plan "perfect" buildings, but should try to make them more amenable to change by the inhabitants, might seem like a flagrant cop-out unless one shares the feeling of the behavioral scientists that the urge to spontaneous expression is also a fundamental human trait. As a consequence, this chapter reviews some of the basic concepts of the behavioral sciences in order to lay a foundation for later proposals. I wish I could write about this material in a way that would make it as fascinating to the reader as it is to me, but if it proves to be tedious, take comfort in the fact that it is probably good for you.

For an architect to attempt to summarize the enormous body of data about human beings that has been accumulated by the behavioral sciences would seem a foolhardy venture. To attempt an encyclopedic summary would not only be foolhardy, it would be idiotic; what is more to the point, it would also be totally unnecessary in a book dealing with environmental design. What is necessary, however, is a discussion of some of the primary points concerning behavior that seem to be most relevant to the design of environments and a comparison of those concepts with the traditional attitudes of the planning professions. Hacking such an enormous subject down to size necessarily requires a rather crude technique that will seem both cavalier and arbitrary to the behavioral scientists themselves, though it should be a service to the designer, the client, or anyone else who is seriously concerned with improving the human habitat.

An initial outrage, of course, is my tendency to lump such proud and independent disciplines as sociologists, psychologists, and anthropologists into all-purpose categories as "Behavioral Scientists" or "Human Scientists". This practice not only obscures the distinctions between these traditional academic departments, it tends to ignore the emergence of a whole set of hyphenated disciplines, social-psychologists, psychological-ecologists, perceptual-psychologists and the even newer families of architectural-psychologists and anthropologist-architects. From the standpoint of an interested observer however, the practice seems defensible. All of these disciplines are focused on some aspect of individual or social behavior and while the distinctions between them may be clear in the minds of the practitioners, they become rather blurred to the observer. Whatever is lost in accuracy by failing to distinguish specifically between these fields seems to be more than compensated for by convenience.

A second factor that is sure to invite some dissent, if not acid criticism, is my choice of certain aspects of human behavior as basic to our society and as having a special relevance for design. This is admittedly a perilous undertaking and is done not only with full recognition that dissent is virtually inevitable but with the conviction that, at this stage, it is also highly desirable. The search for the underlying factors that define the nature of the environments that will permit humans to function at their best is not likely to be an easy one. Within certain limits of objectivity, it is a search that deserves all the discussion, dissent and debate it can get.

One widely held attitude that has stood in the way of a general acceptance of the value of behavioral data in directing human affairs is the conviction that the human animal is so irrational, so unpredictable, and so whimsical in choices and actions that it is hopeless to attempt to deal with him on a systematic basis. This view is summed up in Bertrand Russell's ironic comment, "It has been said that man is a rational animal—all my life I have been searching for evidence which would support this." Starting with such a bias, it is easy to understand why there is some reluctance, not alone in the design field, to believe that the human sciences can offer any reliable guides to the complexities of human behavior.

The attitude that humans are too capricious to be predictable originates when someone else declines to adopt a course of action that we have decided is intelligent and sensible. The difficulty stems, of course, from our inability to recognize that what may seem rational to us may seem completely ridiculous to someone else. The designer who conceives what he considers to be a rational solution to a problem, only to see it ignored or misused, may feel some despair about the capriciousness of human kind, a result that has been at least routine enough to lead to the invention of the defensive term "unsympathetic users." He would be better off attempting to understand the point of view of others and to design solutions that are rational in their terms, not his. As Dr. Donald Young, President of the Russell Sage Foundation, has stated, "It has been amply demonstrated that human behavior is not capricious, that it is subject to humanly inherent and external influences, and that accurate observation and measurement well within practical limits of tolerance are possible."

If human behavior seems irrational to designers it is probably because they do not understand the factors that underlie the behavior. Just what these factors might be is a moot point. Various writers have proposed lists that attempt to summarize them. While they vary in detail and phrasing, they seem to be in general agreement about the universality of such considerations as security, identity, relating to a social group, and so on. In his book, *My Name is Legion,* Alexander Leighton proposes a list of essential striving sentiments that seems to be the most useful for our purposes.

1. *Physical security*
2. *Sexual satisfaction*
3. *The expression of hostility*
4. *The expression of love*
5. *The securing of love*
6. *The securing of recognition*
7. *The expression of spontaneity (called variously positive force, creativity, volition)*
8. *Orientation in terms of one's place in society and the places of others*
9. *The securing and maintaining of membership in a definite human group*
10. *A sense of belonging to a moral order and being right in what one does, being in and of a system of values*

Dr. Leighton stresses that this number is arbitrary. "Given the complexity and intertwined character of human patterns of behavior, the number could be compressed or expanded according to purpose."

They would also have a different priority at different times of life.

The thing that is likely to strike the environmental designer most forcibly about the items on this list is the extent to which they lie beyond his control. This is not altogether unfortunate since it may serve to put to rest some grandiose but baseless assumptions about the power of the environment to alter human nature that seem to float perpetually through the design field. If there is any way to manipulate the environment in order to insure the "securing of love" no one has yet demonstrated this fact convincingly. Another aspect of the list that may seem curious is that neither money nor power, which are so often considered to be potent motivating factors, are included. This omission reflects the fact that they are only effective if they contribute to achieving one of the listed sentiments.

The need for physical security is one sentiment the architect should not have to be prodded about but it is apparent that the intensity of this need is not fully recognized. If it were, there would be fewer instances of design that ignores the apprehension that some people, particularly old people, have about floor to ceiling glass walls and the edges of decks and balconies. Moreover, there would be much more concern about the kind of urban environment that would offer some degree of personal security in walking through the city streets. On the other hand, the architect's only responsibility for the sex life of the people who use his buildings is to recognize that it is a normal human function that must be accommodated. He can't insure that sex will be joyous but he must at least insure that it must be private.

There are other elements on the list that fall into a difficult middleground; they are neither clearly responsive to the environment nor clearly immune to it. The expression of spontaneity or volition would certainly fit into this category. While it is not clear how much the designer could do to promote creativity, it is very clear that he could do a great deal to thwart it. In fact, this would seem to be a point at which the philosophy of the design profession and the tendencies of human beings are in direct conflict. One of the classic tenets in the design field is that it is the designer's responsibility to resolve wholly and in detail the optimum solution for his human clients. To fail to do this is to shirk his professional duties. However, such a totally designed environment leaves very little room for spontaneous expression and certainly does not encourage change. Indeed, with very little effort, the designer can make it virtually immune to change.

It is a little hard to conceive how far the commitment to "total design" can be carried in its most virulent form. In one famous New York office building not only is everything provided for the users and arranged in accordance with a precise plan, but there are also strict rules against bringing any personal furnishings or effects into the building. Just to make certain that there is no doubt about where everything is supposed to be, little markers are neatly stitched into the carpet to insure that desks, chairs, and tables are positioned in accordance with the Grand Design. The result of this policy has been the development of a hot traffic in contraband potted plants and extensive smuggling of bronzed baby shoes, family pictures, bowling trophies and the other indispensable odds and ends we use to personalize our environments. Rules or no rules, during the day the building is used the way people always use buildings; at the end of working hours the personal trivia go back into the desk drawers, the cleaning crews put the furniture back on the markers and all through the empty night the building is a flawless product of total design.

It is only in recent years that there has been any real discussion of the paradox of "total design," and some recognition that in attempting to do a better job, the architects may be thwarting the natural inclinations of the people they are attempting to serve. Such notions, however, are not easily eradicated. The profession is still light years away from a concept that would not only recognize this human tendency but would actually encourage it.

One of Dr. Leighton's essentials, orientation in terms of one's place in society, is one that has been a basic generator of architectural form throughout recorded history and probably before. From the dais in the king's throne room to the paneling in the corporate board room, architectural devices have been employed to announce the status or position of the people who occupy these spaces. In a similar sense the size or height of a corporate headquarters or the innovative character of an institutional building is used to reinforce or to demonstrate the importance of the organization itself. This same concern emerges in the selection of a home or apartment, the name of a subdivision, or the choice of a hotel or restaurant.

So far as those at the top of the ladder are concerned, the design professions have been alert to respond with suitable status trappings. There is no evidence, however, of any general awareness that this concern occurs at all levels. In fact, the general tendency of designers to deal in large overall patterns, to organize and to simplify, tends to thwart the infinitely varied pattern of human ranking. The towering office of a corporate headquarters may announce the importance of the organization itself, but the absolute uniformity of the facade does nothing to reveal the variety of the people within.

Securing and maintaining membership in a definite human group is a sentiment on Dr. Leighton's list that poses some complex problems for the architect. He obviously can contribute nothing to the shared values, background and interests that are the foundation of friendship or membership in a group. He can, however, do a great deal to make contact possible. His decisions in laying out a housing tract, planning the entrance to an apartment, arranging the circulation in a hospital, school, or office, all measurably influence the variety of contacts that are available in the course of the day. It seems very likely that if this factor became a standard element in the design criteria, the opportunity for contact in urban life could be greatly improved.

It is conceivable that as our understanding of the fascinating relationship between life and environment grows we may recognize that the designer can contribute even more support for these essential striving sentiments. Nevertheless, as this brief analysis suggests, there are some positive contributions that can be made now though it may take a realignment of some of the designer's traditional views.

Staying Out of Each Other's Hair

While Dr. Leighton's list of essential striving sentiments is a useful starting point in attempting to understand some of the basic characteristics of human nature, it is necessarily terse. There are a number of other human characteristics that can be cited that are also germane to this discussion of man-environment relations.

One attribute that clearly meets the test of universality is the uniqueness of individual men. In view of rather apparent tendencies that the human animal shows for conforming to the mores and fashions of his group this may seem like a dubious assertion, but it is genetically correct. The odds against two humans, even siblings, being genetically identical are astronomical, something on the order of a trillion to one. To be sure, these same unique humans may attach a great importance to conforming to the standards of their peers, an attribute for which we may be duly thankful since without it community life would be impossible. Nevertheless, human beings are not the same. Even the most consistent and homogeneous group has to recognize individual differences between its members if it hopes to keep them together.

Rene Dubos has pointed out in *So Human an Animal* that the functional significance of man's individuality is not wholly understood, nor is it adequately reflected in our man-made environment. Yet if the concept of "liberty," one of the rocks of our foundation, has any meaning at all, it must include the liberty to be an individual. His quotation from J.B.S. Haldane, the English geneticist, seems to sum it up completely.

> *That society enjoys the greatest amount of liberty in which the greatest number of human genotypes can develop their peculiar abilities. It is generally admitted that liberty demands equality of opportunity. It is not equally realized that it demands a variety of opportunities.*

So long as the design professions continue to work within their present frame of reference, this concept presents enormous difficulties. An architect may be perfectly willing, even eager, to personalize the spaces he creates to satisfy the needs of the individual users but it may be completely beyond his power to do so. In dealing with large housing projects, great office buildings, hotels, schools, and hospitals, he is faced not only with an overwhelming number of individual designs to execute, he can't even identify the individual users. He can only know his user clients in such broad terms as air travellers, hotel users, and apartment renters. While even this general knowledge can be very useful, lacking anything more specific, it does not deal directly with the problem of individuality.

One thing a designer can do to satisfy the urge to individuality is to shape his designs consciously to provide a wide range of options for the user. The ultimate step in this direction would be to abandon the effort to solve all the user's problems and concentrate on an entirely different form of solution that would enable the individual to adapt his environment to his own needs. This concept, which would provide an entirely new frame of reference for architecture and would require a radical re-orientation of professional attitudes, is subject to some very practical limits. Nothing in my own experience would indicate that the human hunger to be unique is so pressing that the individuals in our society feel much compulsion to construct their own environments. Providing them with a stack of lumber and a keg of nails is not the kind of solution they would embrace with enthusiasm. It is the opportunity to adapt, to personalize, to create a better, easier fit between personal habit and physical lair that is important to the individual. Creating the kind of environments that make this not only possible but inviting would be a new assignment for architecture.

Another aspect of individual behavior that seems to be universal, at least in our society, is the desire of people to participate in the decisions that affect them personally. This wish has a great deal to do with the acceptance these decisions are accorded and the enthusiasm with which they are carried out. Recognizing and respecting this desire is one of the basic tools of intelligent management in any field. It should also be a basic consideration in the design field, since it has some important implications for environmental design.

William Michelson of the University of Toronto offers an illuminating example of this factor at work in a survey of Scandinavian planning practices. He reports a study by Bertil Egero that investigated attitudes and expectations of Swedish families both before and after a move into a new community outside Gothenburg. The most striking fact to emerge was that those who had a choice in their new housing experienced the most satisfaction and altered their life style the most while those who were assigned to housing as a result of their position on a slow moving list experienced the least satisfaction in their new quarters and changed their living habits the least. The act of choosing involves a mental and emotional investment to defend the correctness of the choice and to make it work. As Professor Michelson observes, "if the anticipated satisfactions did not materialize, they found others to cite rather than to admit to themselves that their choice of housing had been misguided."

This aspect of individual behavior seems to bear on a point that is a perennial trial to designers: the fact that a solution that is enthusiastically accepted by the group it was designed for becomes a source of growing discontent as the group changes. Much has been made of the ability of human beings to adapt to a wide range of environments, but this doesn't mean that they will adapt cheerfully or willingly to every situation. Their willingness is undoubtedly the result of a complicated set of factors, but participation in the decisions that affect them definitely plays a part. The family that cheerfully agrees to a house plan that requires them to hike up a flight of stairs every time they want to use the bathroom may consider this a very reasonable trade-off for some other benefit while another family forced to live in the house would probably regard this arrangement as the edge of lunacy.

The length to which an individual may go in accepting cheerfully what most of us would consider to be extreme inconvenience is remarkable, so long as it is his own choice. I have a friend, an architect, with strong views about the qualities that he wants in his surroundings. His house is magnificently open to the sun and sky. It is also open, of course, to the heat and glare. As a consequence, he and his family normally wear sun glasses during the day, and so do the friends who are acquainted with the special nature of the house. Wearing sun-glasses is a price he is happy to pay for the sense of space and openness but it seems probable that the tolerance of most families for this much sunshine would be measured in minutes.

The human preference for participating in the decisions that affect our lives presents a complex problem for the designer. To some extent providing options and allowing for the development of self-generated environments would satisfy this preference. The most common method of attempting to deal with it , however, is through a representative committee or advisory group. One of the most encouraging changes of the past decade is the degree to which the public has demanded, and won, participation in the planning process when matters of public interest are at stake. Many federal programs now require the appointment of an advisory group to represent the community viewpoint in the fields of planning, education, health, and welfare.

While the emergence of the advisory group or the representative committee reflects a remarkable shift in official opinion, such groups do not, of themselves, completely answer the problem. As in any group there are complicated questions of group dynamics, personal conflict, and role-playing that influence the decision-making process. What is intended as an agency to mirror public opinion may ultimately produce nothing more than an uneasy truce between conflicting viewpoints. However, after a considerable amount of experience in working with such groups, I believe that their most serious shortcoming is of a somewhat different nature: even under the best of circumstances, and with all the good will in the world, such groups are not capable of reflecting the views of their complicated constituencies through their own resources. Unless they have available to them some means of directly determining these views, they suffer some of the same limitations that an architect or planner would have. They may be of the community but they do not necessarily speak for it.

In view of the restricted representation of "advisory groups", it seems clear that the only recourse the planner has for involving a community, or institution, or corporate body in the planning process is to augment the committee, not with more people, but with the survey techniques of the behavioral sciences. This is not exactly the same as permitting every individual affected by the project to voice his own views, but it is as close as it is possible to come in dealing with large numbers of people. Such a procedure doesn't replace the advisory group, which is essential as a visible agency of representation, but it does give them an accurate picture of the range of opinion within the group they represent.

Survey research may seem like a remote and bloodless form of participation but it is not necessarily viewed that way by the public. In some of the projects where we have used this method, particularly in a university of more than twenty thousand students, it has been gratifying to see that our studies were perceived as offering the student body a chance to participate. Even though only a fraction of their number were actually interviewed, the nature of the effort was widely known and appreciated. Perhaps the individual in mass society has so few opportunities to participate in the events that affect him that even a behavioral survey seems like a step in the right direction.

Before leaving the arena of conflict between the individual and the planning professions, it seems appropriate to touch on one more point that bears on the design process, the individual's need for privacy. The need for privacy is apparently a product of culture rather than an inherently human trait. There are cultures where being alone is not an attractive idea but in our culture it is a factor to be taken into consideration. Being private, or its converse, feeling crowded, is a highly subjective feeling. While we seek and actually revel in crowds at cocktail parties, concerts, and sporting events, the thought of living out our lives in such immediate contact with other human beings would dismay most Americans. Jane Jacobs, with her usual skill in shaping the pithy phrase, sums up this feeling very nicely: "Cities are full of people with whom, from your viewpoint, or mine, or any other individual, a certain degree of contact is useful or enjoyable; but you do not want them in your hair. And they do not want you in theirs, either."

According to Robert Sommer, in his book *Personal Space,* the normal response to unwelcome crowding is "cocooning." Cocooning is the psychological process by which an individual filters out unwanted contacts by ignoring them. The most familiar example of this process at work can be seen in a crowded elevator where a group of total strangers are jammed into an intimate contact they would never tolerate under other circumstances. For the duration of the ride, everyone "cocoons." This, incidentally, is a highly selective defense. In studying street behavior in downtown Los Angeles, I have been amazed at the ability of pedestrians, walking in a zombie-like trance, impervious to crowding, traffic, and the environment in general, to respond instantaneously to stimuli that their selective filter systems were willing to accept, the sight of a friend, a newspaper headline, or a merchandise display.

Cocooning is obviously a useful device for dealing with excessive social contact. It is, however, a device that should not be entirely relied on to satisfy privacy needs. It represents a mental withdrawal from reality, and for a certain percentage of the population who may already have severe withdrawal problems, being forced to cocoon may accelerate their retreat from reality.

Obviously, designers who are involved in planning institutions to house the very old or the mentally disturbed have a pressing commitment to satisfy each individual's need for physical privacy in order to prevent a reliance on psychological withdrawal. The extent to which he should undertake a similar commitment in other kinds of projects is not at all clear. In housing, some of the stress of family life may be eased if we each have an ultimate retreat but there is some evidence that special areas which have been set aside to provide retreats in buildings and in parks are not used to any great extent. In order to have a sense of privacy, there must be some assurance that your privacy will not be invaded. Any retreat that is open to the public obviously fails in that regard. Perhaps the most a designer can hope to do is to provide us with the options that will make it possible, in Jane Jacobs' terms, to stay out of each other's hair.

A Gregarious Breed

To state that the human sciences find the human being to be a highly social animal is, on the surface, hardly newsworthy. Anyone who has given even a fleeting thought to the human propensity for grouping together in huge urban complexes or has considered the implications of our limitless curiosity about what the rest of the world is doing could arrive at the same conclusion without benefit of any advanced degrees. With relatively few exceptions, we all have a strong drive to know and to be known, to see and to be seen, and to have a sense of belonging to a definite group. Alexander Leighton identifies this drive as one of the essential striving sentiments.

In the eyes of some scholars, the facts of man's social nature go considerably beyond such a simple assertion. They regard being social as an integral part of being human. In the view of Rene Dubos, "man evolved as a social animal, and he can neither develop normally nor long function successfully except in association with other human beings."

In this sense, a gregarious nature is not simply a peculiar idiosyncracy of the human race; it is a fundamental part of humanness. Some sense of the urgency of this need can be expressed by turning the proposition around and considering what life would be like if we were forever shut off from human contact. So necessary is contact to our normal life patterns that such an existence wold be, for most of us, the practical equivalent of not being alive. It is no accident that solitary confinement is widely regarded in our society as one of the most extreme penalties that can be assessed.

While this gregarious tendency is widely understood, even widely exploited for commercial gain, the full implications are hardly reflected in the design of our environment. To be sure, architects have produced a wide range of building types, auditoriums, churches, theaters, and stadiums, to accommodate the needs of assembly. They have also, on occasion, designed facilities for assembly, public plazas and the like, that are rarely used for anything. The failure to provide adequately for our urge to get together is found not at the "macro" scale of formal assembly but at the "micro" scale of every day life.

The most immediate and practical illustration of the usefulness of congregation is that if we can't meet together it is difficult to act together. L.E. White, in a study of social problems in a large housing tract in England, came to the conclusion that lack of a suitable meeting space was a major hurdle in the way of a joint community attack on some of their common problems.

> *It must be remembered that there was no tenants' club or similar organization nor any place where tenants might meet or come together for any purpose whatever and therefore no way in which a community problem of this kind could*

be discussed. It will be readily appreciated that this fundamental weakness, the inability of tenants to meet together, was a main cause of the lack of cooperation between the tenants themselves and the solution of community problems such as those under review in this paper and a contributory cause to their failure to cooperate with the housing authority in these matters.

What Mr. White is talking about is not simply a place where formal assemblies can be called, but a place where tenants are in the habit of getting together for their own informal purposes, a facility that encourages the development of a sense of community. It is the sense of community that is essential for joint action. Where it exists, mustering support for community causes is fairly simple. Where it does not exist, creating a sense of community is a formidable task—as anyone who has ever attempted to rally an amorphous suburban neighborhood for either a charitable drive or a campaign against a new freeway can attest.

In a similar sense, the need to meet for group purposes can be found in institutions and organizations that have nothing to do with social action. The faculty club, the staff lounge, and the employees' cafeteria are facilities that are frequently considered by administrators as "fringe benefits," something provided for the pleasure and convenience of the staff but not germane to the principal goals of the organization itself. Since one of the most valuable assets any organization can have is a staff that shares common knowledge, communicates easily, and has a highly developed network of personal contacts, such fringe benefits may contribute more than is immediately apparent. They provide the only meeting ground that many large organizations afford, a place to exchange gossip and ideas, to trade rides and information, to share problems and solutions, and to create a sense of community. Dismissing them as a rather expensive concession to the employees indicates a lack of understanding of the human need for contact. They are a benefit to the staff, but they are also a benefit to the organization. Any organization that hopes to foster a sense of unity within its staff must make some arrangement for facilitating contact and must do so with a full sense of what is at stake. The employee cafeteria should not be conceived of only as a place where the employee can get a cheap lunch but as a community arena.

It is easy to get an erroneous idea of the nature of contact from the preceding examples. An employees' cafeteria may seat several thousand people and a tenants' club may accommodate hundreds, but gatherings of this size don't normally function as a group. It is imperative that planners and designers recognize that, under normal circumstances, group sizes are rather small. While there is some disagreement concerning the precise number, the literature of the behavioral sciences is in general agreement that somewhere between eight and twelve is the maximum number of people who can work effectively together as a group, communicating easily and experiencing a sense of unity and common purpose. Beyond that point subgroups necessarily begin to emerge. This is a process that doesn't require rigorous scientific demonstration. Anyone who has participated in a large conference, or served on a large committee must have experienced the feeling of frustration that such meetings engender. At some point, when it becomes apparent that you have no real chance to participate, you begin talking to your neighbor and a new group is born. As a general rule, it would appear that two small conference rooms are a lot better than one big one in accommodating the needs of a functioning group.

While groups of eight or more can function well when they are brought together for a common purpose, the informal, self-generated, social groups that might be observed in public places are generally even smaller than this. Robert Sommer describes a study of such groups that strongly supports this thesis. Out of 7,405 groups observed in a wide variety of settings, 71% contained only two individuals, 21% three individuals, 6% contained four, and only 2% five or more. In other words, unless some structured, organized activity is involved, informal groups of more than three people are not common. Even in structured, arranged meetings, the number may be small. In one of our own studies of a college campus, a campus with several hundred formal organizations, 65% of the students who attended any meetings at all met in groups of five or fewer.

In relating these facets of informal human behavior to the conventional settings we find in the urban world, it seems fair to say that the design professions and corporate and institutional administrators alike have tended to ignore them or to misinterpret them. The need for assembly is usually recognized only in the formal terms of a conference room or assembly hall. What is more, these facilities are generally designed to accommodate large groups, which occur infrequently, and are ill-suited to small groups, which occur all the time. It is not a question of challenging the useful social function of large meeting areas, but of recognizing the needs of the much more prevalent small group. If a choice has to be made, it seems clear that the first priority must be assigned to nurturing the small groups that are the natural form of human interaction.

One indication that the normal size of the informal group is sadly misunderstood can be found in the arrangement of furniture in lounge and lobby areas of all kinds. These arrangements reflect an assumption that groups of six or eight people are the norm rather than the exception. When two people want to talk—and the majority of groups consist of two people—they usually have to pre-empt a seating arrangement that was designed to accommodate three or four times as many.

The most common failing of the design professions in responding to the human need for social contact has been to deal with it only in formal terms. There seems to be very little awareness that social contact is seldom limited to areas that are formally designated for such purposes. It goes on continually at the bus stop, in the elevator lobby, in the school corridor, at the drinking fountain, on the stairway, even in the men's room. Wherever people meet is a meeting place, and there is little evidence that this fact is generally understood by the design professions. If it were, it is doubtful that public corridors in hotels, hospitals, schools, offices, or apartment houses would be designed as they are. They may serve admirably as traffic routes, but the fact that they are also primary areas for social contact is rarely reflected in their size, configuration, or the way they are equipped.

The same set of comments about meetings would apply also to a slightly different aspect of social contact, making new friends. While the group contacts that have been described apply to formal groups, assembled for some specific purpose, or the response of acquaintances to a chance encounter, some of the same factors are at work in the haphazard contacts that lead to the formation of friendships. Unless contact is possible, friendship is precluded. There are a variety of situations in which the arrangement of the physical environment acts either to encourage or to discourage such contact. In their book *Social Pressures in Informal Groups,* Festinger, Schachter, and Back come to the conclusion that "friendships will depend upon the occurrence of passive contacts and the patterns and frequency of passive contacts among particular people will depend upon the ecological factors of physical and functional distance."

Taken by itself, that statement can be misleading. Friendship, it is generally agreed, is a result of shared values and interests rather than continued contact. It is possible to work in harmony with a group for a long time without developing any warm friendships. On the other hand, people who do share common interests have no way of discovering their potential friends unless some form of contact occurs. In view of the guarded way we approach new contacts, it is possible to go further and say that commonality is not likely to be discovered unless repeated contacts occur.

The rather obvious principle that people who don't come in contact don't become friends, seems to offer the designer an intriguing new assignment. It is not an assignment to design a new kind of facility labeled "place for people to make friends." The challenge is to identify where contacts occur and how the aspects of the design sustain or support such contacts. It should be obvious that contact is something that can't be forced, only accommodated. An apartment house laundry room, for example, is a place where many contacts occur and, if there is a place to sit down, some people may stay there and talk. (Others will still want to get out as fast as possible.) The designer's responsibility is to recognize that a laundry room is something more than utilitarian and that something more than washing clothes takes place there.

As Jane Jacobs and others have observed, contact is also substantially affected by street layout. The large superblock or the winding roads of the suburban housing tract limits the number of contacts that can be made easily because there are fewer channels for pedestrian travel and less opportunity for chance encounters. It is fairly simple to project this insight to other planning problems: the layout of office floors, the organization of waiting areas in the airport, or the design of an elevator lobby in an apartment tower. The matter of whom an individual elects as a friend is strictly a personal choice, but the number of opportunities he has in the course of a day to meet other people is influenced to a surprising degree by the decisions of the planner.

The examples that have been discussed here do not begin to exhaust the vast mine of information that the behavioral sciences have to offer about the human search for contact or the preferences of individuals. All I have attempted to do is to identify some widely accepted assumptions about the wellsprings of human behavior in order to compare them with attitudes and practices in the field of planning and design. It may seem surprising that there are so few "universals" to offer, but the great majority of all behavioral studies deal with specific cases, not generalities. As a consequence, such studies fit more readily into later chapters, related to detailed solutions, than into this chapter on basics.

A Difference in Viewpoint

In discussing Alexander Leighton's list of essential strivings, and the importance of both individual concerns and social interaction as planning guidelines, the focus has been on specific human characteristics and the implications these characteristics have for environmental design. The relationship between stated behavior and the nature of the response required of the designer is direct. If a group of people living in an apartment complex can develop more effective bonds of cooperation if they have a place where they normally meet, the designer's response is obvious: he must provide such a meeting place. If individuals in our society show a strong preference for participating in affairs that affect their lives, the designer must arrange for participation or representation in the planning process. In either of these cases, once the problem is understood, the appropriate response is not difficult to describe.

There is one further aspect of humanness, however, that is altogether different in nature—the way in which we form our conceptions of the world around us. It does not lead to any specific statements about the form of our environment that can directly guide the designer but it is of crucial importance to anyone who is involved in the process of developing concepts about the environment.

The whole field of interests and activities touched on by the term "perception" is a broad one, ranging from the work of physiologists and neurologists in tracing the intricate mechanisms by which we sense external stimuli to the studies of the psychologists in concept formation. We should probably include the philosophers in the roster of interested participants, since any discussion of perception as a physical phenomenon inevitably merges into philosophical questions about the nature of reality. It is a subject of enormous complexity, hardly a topic that fits easily into a book concerned with practical applications of behavior data in environmental design. There is one aspect of it, however, that seems to be of particular significance to the architect or planner: the manner in which we respond to external stimuli in forming our concepts.

The Swiss psychologist Jean Piaget and his colleague Barbel Inhelder provide an insight into this process that promises to clarify a number of puzzles that perplex architects. Readers who have not encountered Dr. Piaget's theories in one guise or another probably accept the concept of associationism as the basis for forming our picture of the world. This simply means that when we encounter a stimulus of some kind we shape an appropriate response, and when we encounter the same stimulus again, we react in a way we have already learned. If you are bitten by a dog, the dog is a stimulus and getting away is your response. The next time you encounter a dog you rely on what you have learned to keep from being bitten again. Stimulus and response are clearly associated.

In a lifetime of studying how a child develops into an adult, Dr. Piaget has demonstrated that the actual process is infinitely more complicated than associationism would suggest. Any stimulus is filtered through a mental structure of action-schemes, a stored supply of behavior patterns that is accumulated through life, and while the stimulus elicits a response, the nature of the stimulus is perceived in a way that depends on the action schemes of the individual. Learning, consequently, is not solely a process of responding to stimuli, it also requires the modification of the individual's action-schemes. Dr. Piaget calls the modification of internal schemes to more closely fit reality "accommodation."

At the risk of belaboring a point, it seems worthwhile to go back over the idea of accommodation in somewhat more detail by pursuing the case of the biting dog a little further. Your response to the bite suggests a direct association between stimulus and response. As you proceed through life, however, keeping a sharp eye out for dogs, you may ultimately make the discovery that a class of dogs exists that won't bite you. At that point it might be said, in keeping with the idea of association, that you have learned to recognize a different kind of stimulus, non-biting dogs, and have learned a new, non-fearful response. This seems to support the idea that learning is a process of making ever more subtle distinctions between different aspects of the world and learning a richer vocabulary of responses. It is at this point that Piaget introduces a distinction of his own. He observes that it is not possible to make this new accommodation to reality without altering your existing action-scheme.

The significance of Piaget's distinction is not apparent until you consider the case of someone who, in the face of a clearly different set of circumstances, is unable to alter an existing action-scheme. His inability to alter his internal scheme of responses precludes the development of an appropriate response to the new set of conditions. To carry our illustration further, confronted with a friendly, non-biting dog, such a person is unable to respond to this new kind of stimulus with a new, non-fearful response as he would if the two were linked by direct association. In order to make this shift, to accommodate his bias that all dogs bite, he must first alter a behavior pattern he has already learned. If you have ever encountered this particular phenomenon, a rigid, single-action-scheme dogophobe, you will recall how ludicrous it seems that an adult can show such clear signs of distress at the sight of a grovelling, harmless cocker spaniel. Having already made this accommodation yourself, it is mystifying to encounter someone who is incapable of recognizing a stimulus in the same way you do and reacting as you do. The explanation, of course, is that his action-scheme alters the nature of the stimulus. While it is the same dog, it is not the same stimulus.

When we leave the area of physical stimuli and begin to deal with more subtle stimuli, the problem of accommodations becomes more acute. While any human being who does not learn to accommodate to the realities of the physical world is destined to a brief and accident-prone existence, the penalties for failing to accommodate to more subtle realities are not so obvious. It is possible for attitudes and action-schemes to persist even though they are wildly irrelevant to the real world. In some instances they are so embedded as to become fossilized, a particular danger when they are elevated to the level of principles." When an idea is advanced or a policy pursued as a matter of principle, in the design field or any other, it is a good idea to examine it with great care. As Ortega y Gasset has said: "our firmest convictions are apt to be the most suspect—they mark our limitations and our bounds."

The general outline of Dr. Piaget's thinking is expressed in the familiar terms "mind-set" and "block," and it is even more familiar as an everyday experience that we recognize but make no attempt to analyze. Edward T. Hall, the anthropologist, describes it this way:

> *There is a growing accumulation of evidence to indicate that man has no direct contact with experience per se but that there is an intervening set of patterns which channel his senses and his thoughts, causing him to react one way when someone else with different underlying patterns will react as his experience dictates.*

The implications of Piaget's insight for the field of environmental design are considerable. First of all, it disposes of the hard-headed position that anyone who fails to see such a concrete object as a building in the same way I see it is either childishly willful or mentally retarded. Christian Norberg-Schulz calls this complacent assumption that everyone sees the world in the same way "naive realism." The designer who assumes that everyone sees the same values, the same interest, the same merit that he sees, is guilty of it. Furthermore, unless he is aware that other action-schemes exist in the minds of his clients he may experience some bitter disappointments. Even more important is the realization that the action schemes in the designer's mind are very special ones; they are formed by training and by the values of his professional group and thus may need considerable revision, may need to be "accommodated" to the viewpoint of the rest of the world. The essentials of human behavior that have been compared with some of the attitudes of the design professions in this chapter strongly suggest that this is the case. The alternate possibility, that human nature will adapt itself to the designer's values, is so utterly unrealistic as to be absurd, yet it is a concept that is implicit in much of the literature of architecture and planning.

One leading west coast architect has been quoted in the architectural press as saying that "good architecture is impossible until there is a change in the public attitude about architecture." That statement reveals a mind-set that is hopelessly irrelevant to the real world. It is something like a doctor saying that it is impossible to practice medicine well until the human physique is radically revised. No doubt surgery would be greatly simplified if we all had a zipper in our sides so that internal inspections and rearrangements could be made more efficiently, but no doctor who still maintained touch with reality would argue that this was essential to proper treatment. The doctor works with the human body as he finds it; the designer must work with human nature as he finds it.

Two factors to which the design professions must clearly adapt are man's desire to see himself as an individual and his strong drive to participate in the decisions that affect him. Current design practices make very few concessions to either one of these characteristics.

While designers are accustomed to working with committees and advisory boards, as a group they have not shown any indication that they see this participation as being vital to the success of the solution. Certainly they have not been aggressive in developing and using the techniques for community participation that the behavioral science survey methods can provide. The designer's tendency to deal in standardized, repetitive units of ten classrooms, one hundred apartments, or one thousand hotel rooms clearly does not conform to the human sense of individuality, and his assumption that he has an obligation to define each of these environments in specific terms, to exercise "total design" control, is also in conflict. Since he can hardly design individual apartments for one hundred families he has never met, he will have to alter his action-scheme; instead of seeking to design a perfect apartment, he will have to design an apartment that the user can perfect.

A second major change in the designer's role must come through recognizing that the purpose of planning or design is not to create a physical artifact, but a setting for human behavior. This will be an extremely difficult accommodation and it is very doubtful that it can be completely made by my generation of practitioners. There are many economic, technical, and aesthetic considerations that shape the buildings we know; they in turn shape the behavior patterns of the people who use them. To reverse this relationship, to start from an understanding of human motivation and let this concern shape the form, will require a profound alteration in the basic approach to design. The patterns of standardized construction components are clearly ordered and consistent; the patterns of human affairs are seemingly chaotic and inconsistent. Designers will require a new mind set that will recognize that it is human behavior that is rational and that the structure that does not accommodate it is irrational.

A last accomodation must come in the realization that training and colleague pressure do engender special action-schemes in the mind of the designer and planner. He is very likely to view these as superior values that the public will, in time, learn to appreciate. The fact that they may not be relevant to the real world will be very hard to accept. A close friend of mine has had a recent experience that illustrates this problem very well. Besides being a talented designer, he is also a warm and considerate human being with a deep concern for the social problems of our day. Given an assignment to design a low-income housing development, he worked very closely with the professional staff of a local renewal agency to create a design that met all the criteria of cost and utility and exhibited exceptional design merit. In architectural terms the units are simple and elegant.

Not far away from this low-income development is an area of the city where a group of superannuated mansions have recently been replaced by new upper income apartment buildings. I hope this latter group represents a class of buildings that is peculiar to southern California, but I am afraid they can be found in many other regions. Again using the terms of architecture, they are ostentatious and vulgar. I think it is fair to say that an architectural jury visiting both sites would be unanimous in finding the low-income project to be outstanding in a design sense and the upper-income housing to be a disaster. A victory of the have-nots over the haves. Unfortunately, the have-nots didn't view their situation in this light at all. They saw it as a case of outrageous discrimination. Instead of blending into the neighborhood, the design of their housing was so distinctive as to mark them as a special caste, a fact that they deeply resented. Their dissatisfaction was so intense that they petitioned the agency to alter their buildings to look more like those of their neighbors.

Obviously this was a traumatic experience for the architect. In reviewing the course of events, he came to the conclusion that his real mistake had been in not insisting on the right to work directly with the tenant group instead of relying on the agency staff to represent them. On the surface this seems like a mental accommodation to reality, a real step forward. His next words exploded that fond dream. His reason for wanting to work directly with the tenants had nothing to do with their views; he felt confident that in face to face sessions he could have persuaded them of the virtues of his design solution. The idea that he might have appropriately made some accommodation during this process never entered his mind.

The kind of mind-set that is illustrated by this episode is by no means confined to the field of environmental design. During a dinner table conversation shared with Margaret Mead, the anthropologist, and a partner in one of the west's leading law firms, I described this incident and got two radically different responses. The attorney saw it as a conflict between superior and inferior values. In his view the low-income families needed expert guidance in raising their standards of taste and should not have had the right to decide these matters for themselves. Dr. Mead's response was more succinct: who says people don't have the right to be vulgar?

This incident touches on a matter of particular concern to the design fraternity, the compromise of design principles. For my friend to satisfy his low-income clients with a building that was, in his view, vulgar and ostentatious, would be unthinkable, as it would be for most architects. What I am trying to suggest is that, before we pledge unswerving loyalty to a set of principles, we should carefully consider how valid those principles are. The aesthetic standards an architect sets for himself are his own affair. How well he serves the community should be a matter of the widest professional concern. In view of the issues discussed in this chapter, it seems reasonable to

propose a somewhat different morality: that the designer's first obligation is to design environments that permit human beings to realize
their potential effectiveness. That obligation is the real measure of
ethical performance, not form or surface appearance.

Chapter III

How Our Surroundings Affect Our Actions

Whenever you enter a crowded bank, post office, super-market, or airport lobby with urgent business to take care of you are likely to encounter one of the universal frustrations of urban life, the queuing problem. No matter how expert you become in sizing up the contents of shopping carts, or estimating the capacity for delay of the gentleman with the bulging briefcase, you are faced with the depressing certainty that the odds are against your getting into the fastest line. All the lines aren't going to move at the same speed. Stress One. If the line you select stalls due to some protracted negotiations about changing planes in Dallas or the advertised special on floor wax, while the others move promptly, another layer of exasperation is added. Stress Two. If some late-comer, using the "meeting-old-friends" ploy or the "dropped coin" gambit imperceptibly merges into the line ahead of you, your stomach is apt to tighten another notch and you have laid a good foundation for a foul mood. Stress Three.

This kind of experience must vex countless thousands of people daily—not enough to provoke a general revolt and not enough, apparently, to bring about a widespread adoption of better queuing methods. It is a perfect example of how one aspect of an architectural setting, the configuration or arrangement, can influence our feelings and our relationship with other people. Given our basic assumption that the rational goal of design is to make people more effective and to harass them as little as possible, it follows that the design of many queuing spaces is far from perfect.

Customer queuing problems illustrate a curious myopia that afflicts the design of many public facilities. In almost any modern setting of this kind it is a certainty that the other side of the counter (or whatever barrier separates public from staff) has received a great deal of study in order to provide an efficient work station for the employee. For obvious reasons, the number of customers any teller or cashier can handle per hour is a pressing concern for management. It is probable that the public side of the counter has also received a great deal of thought as far as appearance is concerned since this is one of the criteria by which a designer measures his own accomplishment. The fact that the designer and his client fail to recognize and accommodate the needs of the people on the other side of the counter is not the result of a cavalier disdain for the customer but because of a limited understanding of the purpose the design is supposed to serve. Oddly enough, this myopia seems to be shared by the public, for while they may harbor all kinds of evil thoughts about the line-jumper in a queue, they are not likely to identify the designer as the real source of their trouble.

Queuing strains are a minor problem in urban life, but they do serve to illustrate the countless ways in which the details of our man-made environment interpose unnecessary obstacles in the way of our individual or group effectiveness. While Chapter II dealt with basic human characteristics that are not adequately served by conventional design methods, this chapter will deal with the arrangement of our physical surroundings that affect both the opportunities and limitations we experience in community life. It can be characterized as a discussion of our immediate environment as a prosthetic device.

The idea that our immediate environment serves the purpose of a prosthesis (as advanced by Constance Perin) requires some accommodation in our traditional views. Putting the arrangement of the spaces where we live and work in the same category with false teeth and artificial limbs is not, on the surface, a rational classification. Yet, in some ways, it is quite obvious. For the amputee entering a multi-story building, the elevator is clearly a prosthetic device. It is as essential to his movement within the building as his wheelchair or his artificial limb. In the same sense, ramps in place of stairways and automatic doors in place of the conventional type provide him with a degree of mobility that would otherwise be completely denied. The components of the building may not be physically attached to his person but they act to extend his capabilities.

The recognition that the details of the environment have special, sometimes poignant, consequences for part of our population has led to a growing concern with "barrier free" architecture. It relates directly to our earlier description of man's anatomical and physiological dimensions. For the elderly, the infirm, the handicapped, and for the very small child, it is perfectly reasonable to regard the physical elements of the environment as being prosthetic in nature, (a view of the artificial environment that places it in the category of media as discussed by Marshall McLuhan).

Having established this relationship, it is not difficult to see that the concept of structure as a prosthetic device applies to man's social and psychological dimensions as well as his anatomical and physiological dimensions. Since the form of the spaces we use, both inside and outside of buildings, has a direct bearing on our personal competence—either supporting or inhibiting our effectiveness as human beings—this view is even more important to the general population than our commendable concern with physical problems. In a social and psychological sense, the settings we use help us or hinder us in three major ways:

They influence the stress we experience in accomplishing our group or personal goals.

They influence the form and nature of our social contacts.

They influence our feelings of identity and self-worth.

That may seem like a pretentious list but the evidence supporting it is very clear. To be sure, healthy, mature people don't suffer psychic collapse because their rooms are badly arranged. They adapt and adjust and accomplish their goals in spite of environmental handicaps. But adapting has its costs in time, frustration, nervous energy, and misdirected effort. The designer who ignores the set of factors listed above is putting unnecessary handicaps in the way of the people who use the settings he creates.

The Daily Stress of System Strains

Of the three influences listed above, the easiest to discuss are the stresses we experience in trying to accomplish whatever goals or objectives we have set for ourselves. They result when the facilities we use make it difficult to do the jobs we have undertaken. This immediately brings to mind a host of commonplace annoyances such as drafty air conditioning, poor lighting, cramped offices, inadequate storage, and so on, but that is not exactly the point. These particular slings and arrows of outrageous fortune you can take arms against. The type of strain I have in mind is the kind you may not recognize, or have come to accept as inevitable. The queuing problem mentioned earlier would fit in this category. Most shoppers would probably assume that getting into the slow line is an inevitable annoyance that no one can do anything about. The obvious idea that a master queue or the "take a number" system would at least insure them an even chance with everyone else might not occur to them. They have too many other things in mind.

The classic example of a system strain is the case of the executive who is assigned a post near the entrance in order to supervise his staff efficiently and as a result spends a good part of his time giving directions to strangers who wander in. I am strongly inclined to believe that this example is apocryphal, invented by someone who needed an illustration for a freshman lecture. I have yet to meet an executive of any rank who would not immediately deploy a barricade of assistants to protect himself in such a situation. Apocryphal or not, I have seen something very similar in studying the operations of a public library system. For reasons which are presumably clear to librarians but are not at all clear to me, the branch libraries in this system were arranged so that anyone who wanted anything had to come to the main desk to ask for it. As a consequence, the trained staff spent a portion of their time directing people to things that could easily have been made self-evident: the location of the card catalogue, the location of the dictionary, and the location of the toilets. The plan may have reflected someone's concept of library functions but it clearly failed to recognize the universality of human functions.

You may not rate this waste of staff time as much of a problem. The staff in these libraries didn't. They were not even conscious that they were performing a job that could easily have been handled by a slight rearrangement and some good signs. They accepted this inconvenience, if they recognized it at all, as part of the system. The same thing is probably true of the library patrons. Having to seek help in finding things that should have been self-evident wasted the patrons' time. If the librarians were busy it meant waiting in line. If the line was too long, it meant doing without. Both patrons and staff suffered some limitation of their effectiveness as a result of the arrangement of the facilities. The fact that it was a minor, even subliminal, strain is beside the point. What is important is that it was totally unnecessary.

Once a system strain has been identified it seems so obvious that it is hard to understand why someone else hasn't seen it and done something about it. The answer is that once an irrational arrangement has been used and accepted, it disappears from view. Having submerged, it is not likely to be identified by the individuals who are directly affected. Only an uninvolved observer, employing a technique of systematic observation, is in a position to isolate the irrational from the rational, and even then it may take a long time.

Another of our studies, undertaken as part of a behavior analysis prior to the design of an addition to an existing school, turned up a classic example of system strain. After many hours of observing the school in use we identified a number of behavior factors that suggested a radical change in the nature of the new facilities. The most curious of these had to do with the process of moving classes in and out of the building. Every morning the children were marshalled on the playground by their teachers and marched into the buildings and down the corridors to their classrooms. At recess the process was reversed and the children were marched out to the playground. With the lunch break and a physical education period, the classes moved back and forth eight times a day, with each little parade taking about five minutes. When an observer finds intelligent people engaging in what seems to be pointless behavior, there is a strong predisposition to assume that it must be serving some worthy purpose that is not apparent, but nothing we could find justified this set of maneuvers. The possibility that this procedure originated in the need to keep the corridors quiet was not persuasive because it didn't keep them quiet. Keeping a group of healthy, exuberant, youngsters quiet on the way to the playground is not only frustrating but well-nigh impossible.

Solving the classroom parade problem was hardly a challenge. Providing outside doors in each classroom so that the class could move directly to and from the playground not only saved about thirty minutes of class time each day, it also relieved the teacher of a frustrating responsibility. What is striking about this case is that in the extensive interviewing we conducted with teachers, administrators, pupils, and parents, no one mentioned that class movements were a senseless waste of time. When it was called to their attention some teachers even defended it as good training—as in some measure it might be. But not forty times a week.

The number of people who are engaged in equally pointless routines in the course of their daily activities is not known, but based on our own experience it seems safe to say they are legion. And they are not limited to the small-scale activities of a specific group or organization. The suburban housewife who is required to load herself and her children into a car everytime she needs to shop for a few groceries is experiencing a strain resulting from someone's decision that it is a violation of good planning principles to permit small neighborhood retail shops to be built within an area of single family homes. If this same housewife feels some anxiety about the safety of her children who are playing in the street, it can be attributed to the fact that street play, which seems to be a universal part of growing up in this country, is not usually considered one of the problems a planner has to solve in designing a residential neighborhood.

It can be argued that large retail centers are more efficient than small ones and that children should play in the parks rather than the streets. Both of these points of view illustrate the widespread tendency to employ formal values in determining planning goals rather than the personal values of the users. Given a choice, it seems probable that American parents would opt for making small purchases around the corner and for streets that are safe to play in. Neither one of the goals would be particularly difficult for the planner to achieve once he accepted them as a mandatory part of his problem.

It may not be immediately apparent that the strain examples described here are really related to human behavior. They seem to fall more easily into the category of annoying misfunctions, the result of a failure to produce buildings and communities that function well. Yet in each of these cases the burdensome actions of the individuals involved were a direct result of physical arrangements. In each instance, they were diverted from their primary goal in order to perform a routine that could easily have been eliminated. It is not at all necessary, as I see it, that the teacher who was spared her daily parade assignment turn that time to good account in the classroom. That would be one useful result, of course, but how she uses the time for the benefit of the class should be of her own choosing. The important point is that she should not be impelled by some designer's oversight to spend her time in an unproductive way whether she likes it or not.

Getting Together

In Chapter II a great deal of stress was laid on the inclination of the human species to seek social contact. This desire to come together is such a basic drive that it is not easily thwarted, even by formal barriers, but it can certainly be hampered by the arrangement of the settings where contact occurs. Two factors that influence our chances of meeting, and the degree of satisfaction we derive from the meeting are proximity and configuration. If we start out close together the chances that our paths will cross are statistically improved. If the point of encounter is arranged so that we can converse in a natural way we are likely to have a more useful exchange.

The influence of proximity in large scale is self evident. It is a matter of geometry. It is axiomatic that people living in Syracuse will know more people in their home town than they know in Spokane. It is probable that they will also know more people on their side of town than on the other side, and more in their own neighborhood than in any other. Sharing the same streets, schools, and shops increases the chances that encounters between individuals with similar interests, backgrounds and values will occur.

It may not be equally clear that the fine scale details of building design and site planning can have exactly the same influence on contact as distance. Two families living on different floors of the same apartment house may be physically separated only by the thickness of the floor construction, but if they use different entrances or use the elevators on a different schedule they are separated in a social sense quite as effectively as though they lived in separate buildings. The same thing applies to the houses in a subdivision. While the houses may be equidistant from side to side and from back to back, if the arena of contact is the street in front of the house, as it usually is, and there is a fence across the back, the neighbors to the rear will not be nearly so "close" as those across the street. In other words, it is *functional* distance that is crucial rather than physical distance.

The concept of functional distance must be amplified to take into account the fact that it is a measure established solely by the user. As a consequence, the designer who undertakes to arrange his solutions so that social contact is easy must attempt to think in terms of user logic rather than designer logic. The two are not necessarily the same.

The influence of proximity on social contact is so obvious at the scale of towns and cities that it hardly needs elaboration, but the extent of its influence at the smaller scale of single buildings is something of a surprise. Robert Priest and Jack Sawyer, writing in the *American Journal of Sociology,* came to the conclusion that even distances of a few feet could have a distinct effect on contact. They were studying the small world of a college domitory at the University of Chicago, a unit so compact and well-defined that it might have been assumed that distance was hardly a factor, but this was not the case.

> *In an analysis of friendship relationships ... closeness or proximity was found to be precisely correlated to recognition and liking. Roommates were both recognized and liked more than floor mates, floor mates more than men in the same house, and house mates more than men in the same tower.*

What is startling about their figures is the dramatic differences resulting from even minor changes in distance. Roommates recognized 96% of their roommates, 66% of those in adjoining rooms, 52% of those two to seven rooms away, and 46% of those eight to thirteen rooms away. The extent to which these students liked as well as recognized their fellows followed the same curve, ranging from a high of 92% to a low of 47%. This particular set of men all lived on the same floor, shared the same baths, the same lounge, and the same elevator. The configuration of the floor was such that the most distant rooms were only thirty seconds apart at a normal walk. These figures are worth a moment of careful thought. The fact that a thirty second walk could exert such a marked influence on knowing and liking is astonishing. As an incredible statistic it is only surpassed by the fact that 4% of the students didn't even recognize their own roommates!

The Priest and Sawyer study went on to measure other social

factors which indicate that while we tend to recognize and like the people we are close to, friendship is a somewhat different matter. When these students were sharing the same classes, the same problems and the same pressures, proximity was a reliable indicator of friendship. Where they did not share mutual interests a different set of factors came into play: peership, the desire to seek out people of like background and shared interests, became paramount. A graduate student living on a floor with undergraduates would seek the company of other graduate students in preference to his closest neighbors, even when it was inconvenient to do so. Seeking friends on the basis of common values, shared experience, or common problems is, as Alexander Leighton has suggested, a normal part of human nature.

Based on this information, the influence of a physical setting on our perennial preoccupation with knowing and being known seems clear. In the formation of friendships the designer plays no part, but in making contact, the necessary prelude to making friends, the designer can do quite a bit. Once the concept of functional distance is understood, he can draw on a considerable arsenal of design devices to increase the probability that chance encounters will occur, not only in the formal settings design usually deals with but in the myriad informal settings where so much of life occurs: parking lots, bus stops, elevator lobbies, laundry rooms, supermarkets and gas stations.

The potential benefit to humankind that might result from a better arrangement of human settings is greater than it may appear to be. If contact leads to recognition, recognition to liking, liking to friendship, and friendship to social activities, then it can be said that contact leads to happiness. On the surface this may appear to be a questionable chain of relationships leading to a preposterous conclusion. "Happiness" is an elusive quality, and ill-defined at best. To imply that it is in any way a result of the planned environment seems to be claiming far too much. Yet Derek Phillips' study of this elusive quality, reported in the *American Journal of Sociology,* found that it was correlated with social activity.

> *The degree to which people perceive themselves as being very happy in contrast to being not too happy is closely related to the extent and degree of their social activities. In short, those people who could report a higher incidence of visits among friends, a higher incidence of people known in the neighborhood, and a higher incidence of organization membership, were consistently reporting higher on the "happiness scale" than those who did not enjoy such activities.*

To draw the conclusion from this evidence that the design professions have some responsibility for making everyone happy would be nonsensical. Too many complex factors are involved to permit any such fantasy. The design professions have a considerable influence on when and where people meet, but only that. Even so, that is a substantial responsibility.

If proximity is the crucial factor in determining the group of people we are most likely to meet, configuration is the crucial factor in determining how well the meeting comes off. This is true not only of the chance encounters we have been describing but also of more formal meetings. The problems involved in conducting a committee session on the down escalator are all too obvious. As a meeting place it offers a near-maximum number of drawbacks. The deficiencies of most meeting places are, of course, not nearly so dramatic, but in any setting *designed* for meetings there should be none at all. With all the information that has been made available by students of interpersonal behavior, there is no excuse for failure in this regard. Edward T. Hall has prepared a table of communication distances between individuals in our society that covers the range from the few inches appropriate for transmitting top secret gossip to the number of feet that constitutes conversational range. Robert Sommer has studied the details of communication by observing how groups array themselves when they are free to choose their own configuration. The choice of seating positions, the angle between participants, and the head to head distance between conversationalists have all been meticulously recorded. The way in which certain seating positions can be institutionalized so that the authority of the position passes to whoever occupies the chair has been noted, and the "Steinzor Effect," which influences the sequence in which the participants speak, has been described in detail. All of these details of interpersonal behavior are directly related to the configuration of a meeting place. They all are, or should be, part of the criteria for designing places where people come together.

The most conspicuous failures in providing appropriate configurations for meetings are not to be seen in the lobbies, lounges and conference rooms designated for that purpose, but in the multitude of contact points where meeting is not consciously considered to be part of the behavior pattern. Meetings can occur in places like stairways and corridors, which are not normally categorized as meeting places; they can also occur in other places, like classrooms, that definitely are meeting places, though they are usually considered as something else. Classrooms, in fact, are an excellent example of how easily the influence of configuration on interpersonal behavior can be overlooked.

F. S. Sumley and S. W. Calhoon conducted an experiment designed to test the ability of school children to remember word groups that were read to them in the classroom. It was not their intention to measure environmental factors in making this study; they were primarily concerned with age, memory span, and word length. The realization that the distance between teacher and pupil, even in the restricted area of a normal classroom, was a significant factor in the child's ability to remember was an unanticipated side effect. The data suggested that an arrangement that put more of the children closer to the teacher would probably lead to better performance in this kind of learning task.

Studies at the college level led to a similar conclusion. In these instances the distance between student and professor proved to be a reliable indicator of the student's participation in class discussions. It might be assumed that the most vocal and confident students automatically chose a position closer to their instructor, but this was not the case. Even when the students were arranged in alphabetical order, distance influenced participation.

From the design standpoint the reasons for this phenomenon are unimportant. So long as this relationship exists, so long as the distance between pupil and teacher has a bearing on participation and learning, it seems reasonable to expect the designer to develop a configuration, circular, fan-shaped, or semicircular, that would put a larger number of pupils closer to the teacher. Unfortunately there is no evidence that this is happening. Almost forty years after the Sumley-Calhoon study was made, the classrooms of this country with few exceptions, continue to be relentlessly rectangular.

Granting that a classroom is used for things other than lectures and granting also that such considerations as construction economics play a large part in determining building form, the fact remains that unless the designer begins his work with a clear idea of the influence of configuration on the interpersonal relations within the classroom, the possibility that he might provide a good solution, more or less by accident, is extremely remote.

The clearest example I know of the influence of configuration on the success people experience in meeting in small groups resulted from one of our own studies. In the course of planning a small plaza in downtown Los Angeles, we conducted a study of public behavior in the streets and parks of the central business district so that we could make the plaza as attractive and useful as possible. All of the parks we observed during this study had some form of seating which was, with one exception, arranged in fixed, formal rows along the walks. Under these circumstances we never saw any seated groups. The groups we observed were always standing. The reasons for this are obvious. Two people seated on a straight bench can carry on a tolerable conversation if they turn toward each other and three people can "make do" if the one in the center leans back and the other two adjust their positions accordingly. Beyond that point, unless some one stands up to face the group, it tends to break up and new groups of two or three are formed. One of the interesting sidelights of this study was the realization that, in such situations, the individual in a wheelchair has some decided advantages. Since he carries his

Park Benches:

a.   The traditional arrangement of benches in long rows
makes it hard to talk to anyone
but the person sitting next to
you. Under these circumstances
the man in the wheel-chair has
the best seat in the house.

b.   When people have a chance, they arrange their seats to suit their natural preference
for face-to-face conversation.

own seat with him, he can face a bench and command the attention of four or five people even though these same people cannot converse effectively among themselves. Once the man in the wheelchair has told his stories and passed on the local park gossip, he can move on to a new audience.

One exception to this group-inhibiting pattern occurred in a park with a small number of movable benches. The benches weren't easy to move; they must have been quite a trial to the older people who used this park, but they had been reorganized into a crazy quilt of open squares, right angles, and benches face to face. Here a man could face his conversational partner and make a point; here someone could "pull up a chair" if he wanted to join the action. The comparison between these two kinds of seating illustrates as well as anything can the disparity between two different worlds, the designer's world of order and system and the human world of change and chance encounter. The orderly pattern of walkway benches produced a configuration that stifled social interchange. The random organization of the movable benches, which would seem at first to have no pattern at all, mirrored with great accuracy the actual pattern of human contact.

The effects of configuration on behavior are not confined to the intimate scale of small groups. They can be seen in a surprising variety of places ranging from the neighborhood sidewalk to the organization of the city itself. Many of these settings have never been studied by the design professions because they are informal and self-generated. It is doubtful that many planners have attempted to deal directly with the phenomenon of teenage "cruising" in laying out new communities or re-planning old ones. Yet this tendency for our highly mobile young people to adopt a certain locale for social display and social contact, for parading their cars and motorcycles and meeting their friends, seems to be a consistent feature of American community life. No doubt many people, particularly the merchants and businessmen in the areas that are elected for cruising, are appalled to find that their street has been elevated to the status of a behavior setting for young people. They would probably wish, most of all, that the teenagers would go away and let them conduct business as usual.

The annoyances that result from cruising scenes are primarily due to an inadequate configuration within the community and its streets. Since few communities make any concession to informal teenage social needs, their needs for display and contact are superimposed on a plan that was intended primarily to accommodate through automobile traffic and the commercial and business needs of the adult community. The result is an abrasive conflict, frustrating for the teenagers and annoying to the adult community. Yet, as Theodore

Goldberg of the University of California at Berkeley has pointed out, the teenagers' needs are relatively simple to satisfy. Essentially they want a place where they can meet in the time and manner of their choosing. Once their needs are described, it is not really hard to conceive of an arrangement that would take care of our mobile youth and also keep them out of the adult community's hair.

Using teenage cruising as an example of how the configuration of a setting influences behavior may seem to strain our definition to the limits. It is certainly far removed from the more usual considerations of a setting as a place formally designed for a specific purpose. What this expansion of the definition actually reveals is that the traditional assumptions guiding planning and design in this country are limited and, what is more to the point, are based on a very imperfect understanding of the elements of human behavior. So long as the designer's proposals are limited to a meager supply of formal solutions and ignore the rich variety of interactions that occur spontaneously there will continue to be serious mismatches between the designed environment and the social needs of the people who use it.

For some reason, the legitimate needs of young people, from infancy on, seem to be consistently overlooked on our planning programs —not in the formal sense, but in terms of their informal activity patterns. To be sure, any adequately equipped housing complex will offer some playground space and most neighborhoods will boast some kind of park. There are formal solutions to youthful needs and while they serve a useful purpose, they do not by any means reflect the realities of youthful life. They require travel, sometimes transport, and always a given block of time. Yet in the life of a child, and even more in the life of the child's parents, such blocks of time are only intermittently available. Jane Jacobs has described the problem in her usual succinct way:

> A lot of outdoor life for children adds up from bits. It happens in a small leftover interval after lunch, while waiting after school wondering what to do and who will turn up, while waiting to be called for supper, in brief intervals between supper and homework or homework and bed. During such time children have and use all manner of ways to exercise and amuse themselves, they slop in puddles, write with chalk, jump rope, roller skate, shoot marbles, trot out their posessions, converse, trade cards, play stoop ball, walk on stilts, decorate soap box scooters, dismember old baby carriages, climb on railings, run up and down. The requisite for any of these varieties of incidental play is not pretentious equipment of any sort but rather space at an immediately convenient and interesting place.

Under these circumstances it is not odd that children often find themselves playing in some place where they are not wanted. If they play in the street they are an annoyance to the motorist and a source of anxiety to their parents. If they play under the living room window they trample the lawn and interfere with the television programs going on inside. Yet if the design of our residential streets were altered even slightly to provide a wide place in the sidewalk much informal activity could be accomodated without annoying anyone.

L.E. White has noticed much the same thing in his study of an English housing complex. In this instance it was the covered entrances or porches that proved to be a focus for youthful activities.

> *It is very doubtful if this very important feature of block dwellings [covered entrances or porches at the foot of common stairs] is sufficiently realized by architects and housing authorities. In wet or cold weather they often become unofficial club rooms; sometimes the children sit on the stairs reading comics; at other times the porch becomes a stage, and, with the stair case and handrail offering some scope for scenery and 'decor', quite elaborate plays, pantomimes, concerts, and dancing displays are given.*

Though the primary purpose of these porches was to provide an entry way to the building, their value to the children could have been multiplied many times if someone had taken the time to "discover" that this was an important setting for them in inclement weather. If this had been done the porches could have easily accommodated both uses without conflict. There is an endless list of instances like this where some aspect of the environment has been arranged, or configured, to serve an ostensible primary function without any thought for the myriad subfunctions it may be required to satisfy. Mr. White's example also illuminates the meaning of the term "functional distance." It is not necessarily the distance to the playground or even to the recreation building, but the distance to the nearest covered area.

The influence of configuration on functional distance and the consequent alteration of movement or behavior has been clearly demonstrated in Richard Myrick's study of high schools in the Washington, D. C., area that has been mentioned before. It is customary in evaluating and comparing school plants to concentrate on the formal aspects of function such as classroom shape and size, educational program, administrative efficiency, and ease of operation. Dr. Myrick's research team undertook to measure a completely different set of factors, the informal patterns of use by the students and the effect of the plan configuration on these patterns. The schools ranged from rather cohesive, compact plans to open-campus types arrangements. In a formal sense they appeared to be quite different; insofar as informal use was concerned they showed striking similarities. In each case there was a primary social center and several minor ones; students "checked in" through these centers several times a day to see what was going on, and the routes students used were a result of social schedules rather than class schedules. Within certain limits, when the arrangement of the school made social contact easier, the student body was more homogeneous. When the school was arranged so that the academic departments were rigidly segregated, the interchange between disciplines that is considered so important in academic circles was seriously limited.

I have no way of knowing the original intent of the planners of these schools or how successful the plans were in terms of classroom arrangement or administrative convenience, but it is evident from the Myrick study that the pattern of student use could not have been a conscious factor in their design. Some schools worked fairly well in this regard but others failed miserably. It is probable that both the successes and the failures were accidental rather than arranged. Yet the student body constitutes an overwhelming majority of the users of any school plant. To plan without their needs or preferences in mind is to ignore the real clients.

One question that I suppose must be considered is whether it is really important to accommodate the special needs of children at play, teenage crusiers, or social high school students. In even the most generous mind there lingers a feeling that young people are, in a sense, undergoing a period of training and would be better occupied in carrying out the assignments of their elders than in their own random activities. This point of view ignores the fact that the social activities of youth are a part of their training and far from being wasteful are essential to maturity. Young people have the same need to know and be known as anyone else, the same need to check their view of the world and the events of the day with their peers. Even in a formal sense this informal interchange can be useful. As Dr. Myrick reported, over half of the informal conversations that took place in the high schools study were related to school activities. Those clusters of conversational students that clog high school corridors between classes are not all wasting time.

The design professions have a great deal to learn about the twin factors of proximity and configuration and their influence on group behavior. It would be more accurate to say that they have to learn to use what is already known. Even more important, they must expand their area of concern outside of the formal settings that are so familiar to them. If the purpose of the artificial environment is to make people more effective in accomplishing their particular goals, then the designer is obligated to respond to this level of need wherever it may be, in city streets, office corridors, bus stops or gas stations. He or she must also accept the definition of need that is determined by user behavior rather than by the sterotypes of design.

Sorting Out the Status Symbols

In discussing the ways in which the physical settings that we inhabit influence our behavior we have, to this point, been describing phenomena that have been the subject of considerable study and discussion. Though the the concept of the built environment as a prosthetic device is relatively new, system strains and the effects of propinquity and configuration have received a substantial amount of attention from the human sciences. In considering how settings influence our sense of identity and our feelings of self-worth, we are entering an entirely different area. While certain aspects of this topic, such as territoriality and the status consequences of our surroundings, have been the subject of much enthusiastic discourse and strong opinions, none of it seems to offer much sustenance for a designer who is seriously attempting to improve the human condition. In one area that I consider particularly interesting, the way in which a building form or design appearance influences our choices, leading us to choose one restaurant to eat in rather than another, there seems to be no solid data at all. Certainly the literature that deals with architecture as an art form has nothing to contribute to our understanding of this intriguing subject. Yet it is a factor that influences our daily behavior in a number of concrete ways.

Territoriality is one characteristic that seems sufficiently well-defined to afford the designer some concrete guidance. It also happens to have received the widest exposure in the public press. Oddly enough, it is possible that more people know about territoriality in terms of nesting robins and baboon troops than they do in terms of human beings. This is undoubtedly due to the fascinating work that has been done in recent years by animal ethologists such as Konrad Lorenz, who are not only capable of painstaking observations of animal behavior but display a remarkable gift of writing about it in the most lucid terms.

While the rules of animal behavior cannot be automatically applied to human behavior, there can be a disquieting similarity between the two. When we see this similarity we tend to classify it from our own perspective as being "very human" when in fact it may be more accurate to see our own actions as being "very animal." In his book *King Solomon's Ring,* Dr. Lorenz offers a delightful example that illustrates this point. Within the flock of jackdaws he was studying was one low-ranking female who had the good fortune to be selected as a mate by a newly dominant male. From the depths of the social order she was elevated in one dizzy moment to the top. In jackdaw society those at the top do not throw their weight around; there is no need for it. They receive all kinds of concessions and priorities as a matter of course and they accept them with the dignity that is appropriate to the responsibility of leadership. This jackdaw Cinderella had no such inhibitions; she hustled for former friends of low rank unmercifully and generally behaved with the utmost vulgarity.

> *You think I humanize the animal? Perhaps you do not know that what we are wont to call "human weakness" is, in reality, nearly always a pre-human factor and one which we have in common with the higher animals? Believe me, I am not mistakenly assigning human properties to animals: on the contrary, I am showing you what an enormous animal inheritance remains in man to this day.*

Territoriality in animals is usually discussed in terms of the defense of a nesting site or a foraging range. It is natural to think of human territoriality in the same terms, a point of view that is supported not only by our concept of personal property, but by an elaborate structure of legal property rights. Even the mildest of homeowners is apt to exhibit a violent territorial reaction if someone dumps their rubbish on his side of the fence. He may even dump his own rubbish on the same spot, but that makes no difference. What he does with his home range is, within accepted limits, his own business. For someone else to despoil his territory in such a blatant way constitutes an intolerable invasion. What is more, the community would be in complete agreement with his point of view. Property rights are not only defended by the structure of laws, they are imbedded in the public consciousness.

Such obvious examples of human territoriality hardly cover the subject. In any large city it is probable that a majority of the people own no real estate of any kind. Yet their territorial urges are just as strong as the suburban homeowner's though they surface in different ways. Their territory may be nothing more than a locker in the dressing room, a desk at the office, or a special seat at a cocktail party but their sense of outrage at an invasion of their private sphere is just as keen as the land baron who finds squatters on the baronial estate. Parents with young children sharing the same room don't have to look far for evidence of territorial behavior. The open warfare that periodically erupts in the children's bedroom is frequently triggered by some territorial transgression.

Robert Sommer, the Chairman of the Department of Psychology at the University of California at Davis, has compiled a substantial catalogue of human territorial behavior that is discussed in his book *Personal Space.* As he demonstrates, territoriality is a behavior factor that is by no means limited to real property concerns. The individual who selects a seat in the library assumes a transitory title to that location that is generally respected. In order to reinforce a claim, he typically deploys an array of personal belongings to mark the extent of his domain—books, briefcases, newspapers, magazines, coats, sweaters, purses, or anything else that will serve to keep an invader out. As a dog uses scent to mark his home range, we use our personal belongings. An array of open books and a jacket slung over the back of a chair are almost as effective in reserving a seat as though the owner were sitting in it. As the Sommer experiments indicate, even a folded newspaper or an out of date magazine serves the same purpose for a limited period of time.

So deeply is this acceptance of individual rights ingrained in our consciousness that we will defend a neighbor's rights even when he is unknown to us. In one of those innocent frauds that are the basis for so many psychological experiments, Dr. Sommer staked out a series of territories in empty rooms using newspapers and magazines. As the room began to fill, his assistants approached the people sitting nearest the stakeouts to ask if these seats were taken. The majority of the respondents reported that, yes, the seats were taken and in some instances even volunteered the information that the owner had just stepped away for a moment. This in spite of the fact that the respondents could not have possibly seen anyone in the seat.

Dr. Sommer's coverage of this subject is so complete that the reader who is interested in the intricacies of human territorial behavior should certainly consult his book.

So far as the design of settings for human activities is concerned, a knowledge of the deeply ingrained sense of territory that is so evident in our culture has several highly practical applications. It can be used both to minimize territorial disputes and to create facilities that are more efficient for a given number of people.

In any of the number of instances in which urban life requires people to live, work, or study cheek by jowl, in offices, dormitory rooms, apartment complexes or suburban subdivisions, the designer can ease some of the stress by being very careful to accommodate territorial concerns. If the tenants in an apartment are assigned a special space in the garage or parking lot for their car and other belongings, the limits of this territory should be very clearly defined. A group of individuals sharing the same office or the same dormitory room should each have a set of facilities arranged so that their personal area is distinct. Even the children sharing a bedroom or the apartment house tenants sharing a common balcony will benefit from clear-cut boundaries. In the course of time, neighbors may make all kinds of mutual accommodations about territory, and the designer's lines may become somewhat blurred, but that is perfectly appropriate. His obligation is limited to providing a framework that works in territorial terms, a configuration that reduces the incidence of territorial conflict. Such a policy won't insure peace on earth, good will toward men, but it should eliminate one source of friction from this earthly paradise.

Territoriality in public places leads to a wasteful use of public

Seating in public places doesn't always conform to the public's seating habits:

a.   A "conversation circle" on the University of Illinois campus. You would need a megaphone to "converse" across this circle.

b.   The group that this circle was probably intended for was a few steps away - in the shade of a tree.

c.   Seating in the Chicago Airport. This was a perfect illustration of the fact that strangers usually sit at opposite ends of a bench But just before I snapped the shutter the gentleman at the far end (with the glasses) rejoined his wife.

facilities. As Dr. Sommer has shown, the student in the library is apt to stake out and protect more territory than he actually needs. Under these circumstances, a library with only half the seats filled may appear "full-up" to someone looking for a place to study. As someone who spends an inordinate amount of time in airport waiting rooms, I have had ample opportunity to observe that much the same thing occurs in the use of public seating. If the first stranger sits at one end of a couch, the second stranger invariably sits at the other end. The third stranger won't normally sit between them. He will head for another couch with an open end. The length of the couch doesn't seem to make much difference. If the first stranger sits in the middle of the couch, it is an obvious ploy to pre-empt the whole thing.

Such territorial behavior offers some rather concrete guidelines for a designer who wants to maximize the use of public facilities. A library table with adequate and very obvious divisions should restrain the sprawlers to some extent and actually increase the useful capacity within the same space. In the same manner, two short couches in an airport lobby will accommodate more people than one long one.

Oscar Newman's study of crime in housing projects described another aspect of territoriality that has extremely important practical applications. Where the housing projects were designed so that the tenants developed territorial feelings about the public spaces they shared with other tenants they assumed some responsibility for these spaces and were able to develop surprisingly effective defenses against criminal activities. Newman attributes this defense largely to the nature of the design: "Design can make it possible for both inhabitants and strangers to perceive that an area is under the undisputed influence of a particular group, that they dictate the activity taking place within it and who its users are to be."

When you enter a new territory and someone graciously asks, "Can I help you?" it may possibly result from pure hospitality, but it is much more likely that you are encountering the first outposts of the local territorial defense network.

The influence of environment on our feelings of identity and self-worth is much more complicated than territoriality alone would account for. Our sense of territory, of the uniqueness of those things and places that we have marked as our own, is a very clear manifestation of our need for identity. The other environmental factors that bear on this point lie in the murky realm of symbolism.

Christian Norberg-Schulz has provided a basis for understanding the role of symbolism as an environmental influence in his book *Intentions in Architecture.* As he describes it, every setting can be characterized as a physical milieu, with rather definite properties that can be defined with some precision. At the same time it can be characterized as a symbol milieu with important properties that are very difficult to define. One example he offers is a doctor's examining room, which must be physically clean as a hygienic matter and must also *appear* clean as a symbolic matter. Indeed, the symbolic cleanliness may be more important for the patient's peace of mind than the actual cleanliness, which he has no way to measure. It requires no great stretch of the imagination to see how this same factor works in a host of related ways. When we select a place to deposit our savings we obviously want an institution that is financially sound, but since we probably can't analyze the balance sheet we may actually base our decision on the symbolic appearance of soundness that is implied by the form and image of the institution. When we select a home or apartment we are not solely guided by the cost, the physical arrangement and the quality of construction; there is a symbolic factor of suitability that plays a crucial part in our decision.

Obviously these are elusive qualities and it is easy to make serious errors in dealing with them. The example of the architect who designed very sophisticated apartments for his low-income clients illustrates the point very well. Obviously, these tenants read the symbolism in a different way than he did. Given a choice, they would have elected the more prosaic forms that identified them with the general community. This same kind of unexpected reaction shocked the architects of an eastern school that turned a windowless facade to the rather bleak surroundings. The neighborhood interpreted this as a sign that something was going on inside that the neighbors weren't supposed to know about. A perfectly innocent design decision became a symbol of rejection.

The effect of architectural symbols on our actions is most apparent when we encounter new settings. A family travelling across the country by car must make a series of decisions about where to eat, where to shop, or where to spend the night. For the most part, their decisions will be based on an interpretation of very subtle symbols. The state of their clothes, the condition of their car, the mood of the children, and the health of their pocketbook are the personal factors that define their needs. These are compared to the symbols displayed by the establishments they are considering, both overt and covert; one looks too expensive, another too shabby, a third too formal. In this process of evaluation the only evidence available is the external appearance of the establishment. If the man and his wife were travelling the same territory alone, or if the man were looking for a place to entertain business associates, the personal factors would be different and a different decision might be made.

A family moving into a new neighborhood goes through a similar process in making initial contacts with the stores and shops they will patronize. In time, of course, once they have learned the neighborhood lore about where to go for these goods or those services, they may elect an entirely different set of sources, but at the outset they must make their decisions on the information they can get from what they see. Making judgments on the basis of external appearances is a notoriously poor way to select anything, particularly eating places, but in a new setting it constitutes the only information available.

The ability to sense what is appropriate, based on clues or symbols that may be virtually subliminal, can be highly developed. In the past few years I have taken to riding a motorcycle to the office and to business appointments in the city. It is highly maneuverable, easy to park, and produces only a fraction of the exhaust emissions of an automobile. Most important of all, it is a lot more fun than driving a car in city traffic. Very early in my motorcycling experience I pulled into a downtown parking structure and was refused admittance. This was an astonishing experience. When you have spent a lifetime with the calm assurance that you will be welcome any place you choose to go, to be flatly rejected by a parking garage passes all comprehension. It was an experience that gave me a faint glimmering of the sour taste of senseless discrimination.

The garage attendant who was on duty is not likely to recall that encounter as one of the happiest moments of his life, but then, neither do I. No rational human being wants to squander his emotional resources in pointless conflict, particularly with someone who has no authority to change the situation. As a consequence, I very quickly learned to identify places where motorcycles were welcome. Within a few weeks, I was able to approach even a new parking lot and make an accurate judgment about the attitude there. I can't tell you what enters into this evaluation, but for several years it has been unerring. There is a very fine distinction in my evaluation. I am sure about the lots that will let me in but I'm not sure about the lots that won't. As a result I probably pass some places that would be perfectly happy to have me turn in. Their image is ambiguous and since I can't interpret it with certainty, I pass on by. Losing the motorcycle trade is not apt to hurt their feelings anyhow.

That is not true of other business establishments or institutions that survive on the basis of public acceptance. For any appreciable number of potential customers to pass them by because they presented a misleading or ambiguous image would be a matter of serious concern, yet it seems very likely that this happens all the time. Since our present knowledge of the symbol milieu is so fragmentary, there are very few guidelines to follow. The businessman makes decisions about the appearance of his quarters, or has an architect make the decisions, without any concrete information about how these forms or design features will be interpreted by the public. For the most part he reflects his own tastes or own judgement, which may not even be remotely like the judgment of his potential customers.

Although the symbol milieu is a mysterious quantity, it is not completely impossible to deal with on a rational basis. Indeed, the first step must be to assume that, for the most part, it is not in the realm of mysticism at all. The reaction that most of us have to the symbolic aspects of a setting must lie in our own needs, interests, and self-evaluation. We hesitate to enter because we fear a rebuff, either social or economic. We pass on by because there are other options available to us that are clearly more suitable to our needs of the moment. If a catalogue is constructed of the possible interests and concerns of the potential users, then some rational forecasts can be made about the design elements that will influence their choices.

A study that we conducted for a banking system in California illustrates this point. Our assignment was to design a prototype branch bank that would embody specific operating features that could be incorporated throughout the system. In evaluating the bank's existing facilities from the public's point of view we found a series of symbolic affronts that had unintentionally been built in.

> *Anyone entering the parking lot was greeted with a blank stone wall.*
>
> *Anyone who arrived before banking hours had to stand at the front door or sit on the sidewalk.*
>
> *Anyone who entered the front door was greeted by an impressive view of the back door (and vice versa).*
>
> *From the entrance it was impossible to tell, because of depth perception problems, which teller windows were open and which had a teller free to take care of a customer.*

At first glance these items may not appear to have anything to do with symbolism, a topic that is usually discussed in terms of the mysterious influence of certain forms and shapes in evoking an emotional response of fear, joy, apprehension, or repose. Architectural forms may have such an influence, though this is certainly not clear. The kind of negative symbols we found in these branch banks are, in contrast, very clear. The handshake you exchange when you meet a stranger isn't a very important symbol, but when that stranger refuses to accept your extended hand, the symbolism is suddenly loaded with meaning. Exactly the same kind of rebuff can be read into the entrance features of the banks we studied.

Someone may ultimately classify the symbolic influence of shapes and forms on our emotions and provide us with a design vocabulary of singular importance, although it seems unlikely. There are too many cultural aspects, too many fashion changes, and too many personal histories to make it probable. Symbolism in terms of normal human relations is something that is much easier to understand. The openness and candor with which we greet a friend, the arrangements and concessions we make for his comfort and convenience, are symbols that are clearly understood. To design settings that express these

A friendly greeting from your friendly, full-service bank:

a.   From the front door you get a great view of the back door. The plan of this bank looks rational but it has some flaws from a customer's standpoint. Due to depth perception problems the entering customer can't tell which teller windows are open or which tellers are available.

b.   Looking directly into the tellers windows made things much easier for the customer. These pictures are part of the study that led to the design of the queuing contraption shown elsewhere.

same courtesies is simply an extension of an accepted pattern of human behavior.

* * * * * * * * *

Defining the influence of settings on human behavior in the three ways that have been discussed offers some substantial benefits in organizing information and analyzing problems. At the same time it introduces the risk of oversimplfying some subtle and intricate relationships. Any setting that makes it difficult for us to function efficiently not only induces some personal stress, it also influences our feeling of self-worth. The manner in which we read symbolism in building forms, which influences the haunts in which we are most comfortable, also influences the social contacts we are apt to make. In short, all these factors are inextricably interwoven in the real world.

Nor can an understanding of the influence of settings be considered complete without reference to the human characteristics defined in Chapter II. The desire to participate in the decisions that affect us obviously bears on our sense of identity yet it is affected more by the process of design than the setting itself. It is more accurate to view the salient characteristics that have been brought out in these two chapters as the identifiable features of a panoramic landscape rather than a total description. The human experience is a richly varied continuum rather than a series of discrete episodes.

Nevertheless, the evidence supports the thesis that the designer who regards the planned environment as a human prosthesis can make a distinct contribution to human effectiveness. It is not likely to be in heroic terms; there is very little chance that he will alter the form of society or control the course of history. His contribution is defined in the much more limited terms of everyday life.

Some years ago I described the concepts outlined in this book to an audience of architects and educators in the Midwest. In the course of the discussion I referred to Edward T. Hall's work in defining the manner in which members of different societies view the bubble of personal space around their bodies. Dr. Hall has coined the term "proxemics" for this phenomenon. He has demonstrated that our society has an essentially non-contact culture. In other words, we don't enjoy having people get very close to us except in special circumstances. After the lecture, a dean at one of the large universities in the area came up to the platform to find out more about this concept.He had never heard of Dr. Hall's work and found in it a clue to a poignant problem of his own.

His sense of personal space was so acute that when it was violated he experienced difficulty in his normal functioning. If he went to the men's toilet at the airport prior to departure and found a line of men at the urinals it was difficult for him to enter. If his needs were such that he forced himself into line, he was unable to function so long as he was close to somebody else. As a result, on several occasions he had to retreat, board his plane, and wait for it to take off before relieving himself. Anyone who has had to wait at the end of a runway while an interminable line of jet aircraft took off can understand his agony.

Some people regard this story as excruciatingly funny. I tend to view it as very sad. I have no idea what the source of his problem might be or what he should do about it. I do know, as an architect, that it is a very simple matter for me to see that stall dividers are placed between the urinals in public toilets. It is not a heroic matter or a heroic measure, but if it makes it somewhat easier for even a few men to maintain their self-respect, their sense of competence, then it seems to be eminently worthwhile. How often do any of us have a chance to contribute so directly to the well-being of another human being?

Chapter IV

Architect And Behavioral Scientist In Joint Operation

In order to develop designs that accommodate the behavior traits of the human species, the architect of a building project or planner of a new community needs information at several different levels. The basic aspects of human motivation embraced in Alexander Leighton's list of essential striving sentiments provide the kind of information he needs to define the purpose of shelter and to re-examine his own professional attitudes. The influence of settings on human behavior, discussed in Chapter III, demonstrates that there are some general principles of configuration that produce more sympathetic environments. While both classes of information are indispensable, they do not deal directly with the designer's principal enigma, the needs of the unique population he is attempting to serve. In order to describe a solution with the precision required for construction, he must have precise data to work with.

At the present time the only hope of assembling the precise information that the designer requires is through some form of research focused on the project area or project population. In view of the kinds of information that are required this inevitably leads to collaboration with someone skilled in the research procedures of the behavioral sciences. Joint operations of this nature are new in the design field and pose some perplexing problems for the designer. The benefits are so obvious, however, that I believe it should become a standard practice. The best way to illustrate the method, the problems, and the potential is to go through an actual case, the development of a behavior program for a new student center at a large urban university.

California State University, Los Angeles, is one of a new crop of colleges that has appeared in this country in the last quarter century in response to an expanding population and an increasing percentage of people who need or want higher education. It serves over 20,000 commuting students on a year around schedule, with classes running from eight in the morning until ten at night. Like many other urban schools it provides no housing for its students; if they are living away from home they must find housing in private facilities.

The student population at CSULA differs from others in several ways. It reflects the growing practice of providing for a true continuing education; the more typical college group in their early twenties mingles with a substantial number of middle-aged men and women who have returned to the campus on a part time basis to obtain additional training in fields that are pertinent to their activities, or to get the degrees that may be necessary to advance in their occupations. As a consequence, to lump this diverse population together under the general heading of "college students" would be misleading.

In the course of its rapid growth the school had acquired a full complement of typical academic structures; classrooms, labs, a gymnasium, athletic field, administration building, and a library. Unlike other schools, however, it had no center for activities of a non-academic nature, the kind of building that is customarily labeled a "student union," "student center," or "campus union." Since it has been the policy of the state not to provide this kind of facility with public funds, the students had voted an annual assessment to finance it themselves.

In preparing to design a building such as a campus union the very first requirement is some description of what purpose it is supposed to serve and what facilities it should provide. This is generally termed the "building program" and is a crucial document for the success of the project. There has been a growing awareness in the architectural profession that the individual or group that prepares the program in effect designs the building. No matter what skills or talents the architect may bring to the project, he is limited by the quality of the program. If it fails to satisfy the needs of the people who are going to use the building there is only a limited amount he can do to ensure the project's success. Another and even more prevalent problem is that if the preparation of the program is dominated by a strong willed individual, and there always seems to be one, it may satisfy his personal priorities very well but do a miserable job for everyone else.

Depending on the attitudes of the university administration the job of compiling the program could have been handled in several different ways. On the premise that students come and go but the university goes on forever, the assignment might have been given to the facilities planning experts on the administration staff who could be counted on to reflect the long range interests of the institution. They would probably, in turn, retain an architect or a specialist in programming to assist them in sorting out the myriad details that any program for a large project involves.

A more democratic approach, and a much more normal one would have been to appoint a campus committee composed of students, faculty and administrators. Since a committee would have a hard time doing the complicated detail work required, they too would probably engage a consultant to develop the details in accordance with their general policy decisions. In either of these cases, a considerable amount of time would probably be spent in visiting similar buildings and discussing similar operations with their counterparts elsewhere. If this group were sensitive and responsible, they might also undertake some kind of public opinion poll on campus through a questionnaire describing the typical alternatives that characterize buildings of this type. Ultimately, they would produce a document defining the requirements for the building, the size, and the budget. This document becomes the architect's bible.

This is a brief description of a planning procedure that has been followed innumerable times, not only on college campuses but in an infinite variety of building types. The only major deviation might be that in many instances the programming effort would not be nearly this well organized and thorough. It might reflect, instead, what are essentially the preferences and priorities of one individual—corporation executive, minister, hospital chief of staff, or land developer.

In spite of the fact that a multitude of buildings have been designed on the basis of such information, the traditional procedure described above exhibits some basic flaws. Studying other buildings is a great way to identify errors that you might otherwise repeat and to pick up useful ideas. Unfortunately, the method also begins to build a stereotype of what a particular class of buildings ought to be like, even though the examples studied may be serving groups with backgrounds, aims, and problems that are entirely different from your own.

Programming by committee also suffers from the limitations of any committee activity. Regardless of how sincere and conscientious the members may be, there is no way in which they can accurately reflect the divergent points of view of all their constituencies. Without some special effort to identify each of these points of view the committee members must fall back on their own personal judgments and these may reflect more of their own values than the values of their constituents. The faculty representative has a hard time escaping the concerns and unique perspective of his own discipline. The full-time student finds it difficult to evaluate the special needs of someone who is working his way through college or the harrassed woman who is trying to get enough credits for a teaching credential while raising a family. In addition, the committee is beset by all the normal stresses that occur in such a group; certain individuals are more aggressive, factions form and trade support, and some members may be muzzled because they rank too low in the hierarchy to challenge the leaders. It is, at best, an imperfect way to summarize the requirements of any large group of people.

In initiating the design of the campus union at Cal State a different approach was taken. There was a representative committee, to be sure, and a staff executive to compile the detailed requirements of the program, but before any decisions were made, a comprehensive behavior study of the entire campus community was undertaken. The premise behind this study was simple: if a building is considered primarily as a behavior setting and is intended to make human beings more effective in pursuing their personal goals, then the proper starting point for any planning venture must be a study of the people who will use the building. Before an appropriate building program can be defined there must be a description of the behavior the building should accommodate.

It is at this point that it is possible to gauge more accutately the usefulness knowing the general human propensities that have been previously discussed. Based on such general knowledge we could say with some assurance that the campus community would want to participate in the decision–making process. If the new building were assumed to be a focal point for social contact and interaction, we could even describe the kind of configuration that would facilitate interaction. Beyond this point we would begin to run into trouble. Our generalized knowledge could only be useful when it was brought to bear on the solution of a specific problem. Unfortunately, nothing in the literature or lore of the behavioral sciences would provide us with the specific information we needed to know about the objectives of this particular group of people. In order to be responsive to their needs we would have to know them in great detail and in all the variety implied by such a complex population. In order for the building to play a supportive role in their lives, we had to know the goals they had set for themselves and the stresses they experienced in meeting these goals.

For an architect to attempt to measure complicated behavioral and interactional factors without special training would be patently ridiculous. The social sciences cannot be mastered intuitively. It would require great skill and experience in the method of the behavioral sciences to organize and conduct a large-scale study and to arrange the data in a usable form. If we hoped to understand the complicated Cal State constituency and plan for it intelligently we could only do it in collaboration with someone who had the appropriate talents and commanded the requisite skills. Only a joint operation that drew on the specialized abilities of the behavioral scientist for measuring human factors and the synthesizing talents of the designer to define the form of the structure could deal with the complications inherent in this approach to architecture.

The Cal State project was not unique in that regard. Other than private dwellings or other structures where the owner is the sole user, I cannot think of any building or planning project involving human beings that would not benefit from this kind of collaborative approach. If an architect is seriously committed to reflecting human behavior in his design solutions, he has no other choice. While it is true that behavioral scientists have assembled an enormous body of literature about the human animal, it is very low-grade ore as far as the designer is concerned. He is always dealing with a specific project in a specific locus and global generalities about human nature offer him very little sustenance. In time this situation may change; something approaching a design manual for human affairs may eventually emerge. Until it does, the architect's best bet is to apply the research methods of the behavioral sciences to the specific project in which he is engaged. In order to do that he has to arrange some form of collaboration or joint operation with an expert in the behavioral field.

In our case, the answer was simple. We had been working with Dr. Thomas Lasswell, Professor of Sociology at the University of Southern California, for a number of years. In the course of developing our particular approach to architecture we had tried several different techniques of collaboration ranging from simple consultation to more complex studies of behavior in architectural settings. As a consequence, we had a well developed set of tools to apply to the Cal State project. Before I describe how these were employed and the results they produced it would be worthwhile to review the options that are available to an architect and a behavioral scientist in the development of a behavioral program for a specific project.

Ways of Working Together

The range of project types and project sizes embraced within the scope of architectural practice in this country is enormous. The size varies from simple alterations and additions to existing structures to vast commercial and institutional complexes. The variety encompasses the range of human activities, from single-family dwellings to complicated centers for the healing arts. If this scope is enlarged to include the planning field, the development of new towns and the renewal of existing communities, the number of people affected by a single project can vary from a relative handful to tens of thousands.

In developing the building programs for such disparate projects it is obvious that the architect or planner has widely disparate resources of time and money to commit to the programming process. In some instances the program will be developed after a hasty conference, figuratively or literally on the back of an envelope. In other cases years of program research may precede the planning of a major project. This is not an argument in favor of careful planning for large projects and casual planning for small ones. On the contrary, it is my own thesis that size has nothing to do with importance. The humanistic view insists that in either case, large or small, affluent or impoverished, the human beings affected by the designer's work deserve the most conscientious study and concerned attention that can be brought to bear on their needs.

Nevertheless, to propose an idealized procedure for developing a behavioral program and insist that it be universally applied would be quixotic. Regardless of what might be theoretically desirable, every architect and planner must ultimately contain his activities within the resources that his clients can or will commit to the project. As a practical matter, he will adapt the scope and scale of his behavioral involvement to whatever is possible.

At the very least, the designer will acquaint himself with the current thought in the human sciences that applies to the man-environment equation. While generalized concepts are of limited use in dealing with specific planning details they at least open the mind to a realization that the conventional attitudes of the design professions show some spectacular weaknesses in generating environments that are sympathetic to humans. With even a limited background, a designer at least acquires the capability of questioning some of his easy assumptions about the purpose his design is intended to serve. Indeed, if he can't make this basic breakthrough in his own mental systems, more elaborate processes won't help him anyway.

Step two toward improved behavioral programming is to establish a working relationship with someone in the human sciences who has some skill in employing the information-gathering techniques of these disciplines and has exhibited an interest in applying these skills to practical problem solving as well as pure research. Not just any Ph.D. will do. It takes someone who feels that the rather mundane problems of the built environment are worth solving, someone who is capable of true collaboration, and who is willing to accept the arduous time and money pressures that are an inevitable aspect of any planning program. Of all these limitations, the time pressures are the most difficult to cope with. In pure research the work is done when you have found the answers you were searching for. In most planning programs, the work is done in accordance with an inflexible master schedule.

Once a contact with a behavioral scientist is established, the designer has opened a channel into a field of information and a set of methods that can be of inestimable value to his clients. At the simplest level he has someone to turn to for verbal consultation and to review and evaluate the possible effects of alternate solutions. In a sense, the behaviorist can play the role of ombudsman for the users, and even if he or she may be severely limited by the lack of specific data, he will almost certainly add a new dimension to the design decision-making process. If time and finances permit he can go on to conduct a literature search in the endless archives of human behavior in order to expand the range of information that can be brought to bear on the solution of a given problem. That is no light undertaking, incidentally. Until you have personally attempted to cope with the *Sociological Abstracts* or the *Psychological Abstracts,* it is hard to comprehend the staggering amount of study that has been devoted to the human animal. It is available in almost any tonnage you require. Wading through these mountains of information in the hope of finding some nugget that will make life easier or more effective for the particular group you are concerned with is a frustrating experience. Nuggets are few and far between.

The behavioral consultant is not really able to get into high gear until he is given a chance to employ some of his more elaborate information gathering procedures. So far as the planning process is concerned, it is these techniques that offer the best hope of improvement rather than broad generalizations or the minutiae of some exotic environment. This is not solely a personal view. Robert Sommer has expressed the same conviction: "I believe that social sciences make their greatest contribution by offering methods by which information about human behavior can be objectively and validly obtained, rather than formulating detailed laws about people's responses to blue walls, round buildings, or thatched roofs."

Once the behavioral scientist is unleashed, so to speak, which is mostly a matter of arranging funding and providing a sufficient time period, he can develop the required behavioral data in whatever detail is necessary. In the simplest terms he uses two techniques in assembling his data, he observes and he asks. On the surface these appear to be elementary, but in practice they are highly complicated and require considerable sophistication if the results are to be valid. (This is especially true of the questionnaire or interview procedure.)

While the value of both observations and interviews are greatly enhanced if they are used together, in a pinch either activity can contribute vital information. An ideal sequence would be to make observations first and use the data collected to determine the kind of questions to be covered during the interviews. For example, if you were planning a neighborhood renewal program or even a neighborhood pizzeria, you would start by observing public behavior in the area. If your plans included some facility the neighborhood did not have you would have to use a surrogate neighborhood, at least so far as that activity was concerned but your primary focus would be the people in your project area. This may strike you as an odd way to embark on a planning program but I assure you from a rather substantial personal experience that it is richly rewarding in the most practical way. It is also apt to raise some intriguing questions that can only be answered by the interview process.

The observation process is one that should be particularly attractive to the designer since he can, with some application, develop a competence himself. He may never acquire the skills or the thoroughness of someone with a behavioral background, but in the absence of a budget for behavioral consultations he can certainly improve the quality of his design decisions by conducting his own observation studies. This kind of self-help system won't match the work of a professional like Robert Sommer, but it is much better than nothing.

Anyone who is interested in undertaking observational studies will certainly profit by a study of Robert Sommer's work. He will also find it helpful to acquaint himself with Roger Barker's concept of behavior settings and Constance Perin's concept of behavior circuits. Both of these writers offer a useful theoretical basis for observation and a framework for organizing observational data and compiling it into a useful form. Systematic studies of this kind are called ethological studies and are an exact parallel to the work of animal ethologists such as Konrad Lorenz and Nikko Tinbergen. The work of such men in the field of animal behavior has demonstrated that observational studies are extemely valuable and have lead to some surprising discoveries about animal life patterns. It is dismaying to realize that relatively little of this kind of systematic observation has been devoted to the human animal.

For the amateur human ethologist the key word is "systematic." He must define a target area, school ground, apartment complex, airport lobby, bus stop, post office, or any other arena of action and study human behavior in this locale in a systematic way. He must chart a route that puts him in touch with every potential behavior setting in the area, and repeat the observations at different times on sufficient occasions so that he can record the full range of behavior. During this process he is recording what he sees and, if possible, is photographing the settings for future reference.

Systematic observation is obviously not the same thing as a pleasant ramble while enjoying the scenery and the ambience. The observer is busy all the time, counting heads, identifying groups, and noting the minutiae of individual behavior. He must avoid the trap of assuming that people are doing what he himself would do under the same circumstances. He must be as conscientious in noting what people are not doing, or what areas they are not using, as he is in following the center of action. Most of all, he must avoid making hasty judgments about the significance of what he sees. The time for judgments is later, after all the data are in hand and after the notes and photographs have been independently analyzed by someone who was not at the scene.

One of the most delicate problems of observation stems from the effect the observer's presence may have on the action he is studying. This is a factor in all forms of behavior study. The "Hawthorne Effect", named after a classical study conducted at the Western Electric plant at Hawthorne, Illinois, may occur. It reflects the fairly obvious fact that being involved in a test procedure or being the object of close scrutiny alters the behavior of the individual being studied. As a consequence, the observer must be as unobtrusive as possible, a status that becomes much easier to sustain as he becomes a familiar part of the setting.

The Hawthorne Effect overwhelmed an early study of mine when, in an effort to understand college domitory life, I lived for a week in a dorm and took all my meals in the dining hall. Disguised in a conservative tweed suit and vest, I was about as inconspicuous as a piebald aardvark. During my tenure, dormitory life proceeded with the ordered calm of a monastery. The only thing I gained from this experience was the cynical suspicion that if the American student is exposed to dangerous influences during college life, the attack is focused on his stomach rather than his head! In time, of course, I might have become as much a part of that scene as the faded wallpaper and at that point the normal life of the dorm could have been studied with some accuracy. It would be a rare circumstance, however, when an architect could devote that much time to such a study. As a consequence, in any instance where he is an obvious intruder, an observer must work through intermediaries who are a normal part of the scene. In this case it would have been natural to have the observations made by students.

Another device that does work on occasion is to assume the role of a nonperson. One of the curious quirks of our territorial behavior is that it is only activated by others that we recognize as people. Telephone repairmen, janitors, mail clerks, and a host of others play special roles that permit them to move with immunity in areas where others would fear to tread. They are not invaders because, in these roles, they are seen as non-persons. I'm not suggesting the use of elaborate disguises in order to be unobtrusive for these are not essential. The disguise I assume most often is that of a rather absent-minded middle aged architect. For some reason this is a role I play very easily. It is, also, a role that has a high invisibility quotient in many public places.

On one occasion, for example, it was necessary to obtain some behavior-setting photographs in a part of the city with a very limited tolerance for pollsters, scholars, public officials, curiosity seekers, or anybody else who might be classified as an observer. Working in that area with a miniature camera would have been risky. Doing the same thing with a tripod and view camera, trudging through the district like a reincarnation of Mathew Brady, didn't provoke anything more than a mild curiosity. A harmless eccentric is clearly a non-threatening non-person.

Once observation data have been assembled and classified it is almost inevitable that questions will emerge that can only be answered through some form of opinion poll or interview program. It is an activity that the amateur is well-advised not to attempt himself. It requires skill and experience to phrase questions that are free of bias and that accurately reflect the subject's personal views rather than the attitudes of the person who prepared the questions. Questionnaires and interview schedules must be tested for validity, and interviewers must be trained in their use. A particularly complicated aspect of this work is the selection of a sample that represents a true cross-section of the group under study.

Once the interviewing is completed or the questionnaire results are in hand, the data must be sorted and analyzed if they are to make any sense at all. Even in a very modest project this virtually demands the use of a computer and the development of appropriate computer programs. The whole undertaking is one that is better left to experts. The architect or planner will find that he has his hands full in trying to figure out what to do with the information once it is delivered to him.

The focus of such a study can obviously vary to fit the need. An organization that planned to move its headquarters might test the suitability of various locations in terms of employee travel time in order to minimize the disruptive effect or the loss of staff that might otherwise result. A land developer or home builder moving into a new area would need to know the income level of the community, family size, and local preferences in housing. A school administrator might invite his teachers to comment on the suitability of various classroom arrangements, storage systems, or teaching aids. Each of these highly focused examples would contribute to intelligent planning by measuring existing attitudes and preferences.

The type of study most often encountered in the design field is the user preference survey. While it is not as common as it should be, it has been used enough to demonstrate its value. It is nothing more than an attempt to elicit from the potential users their own views on the facilities that should be provided in any new project that is being planned. Housewives may be questioned about room arrangements, office employees surveyed on desk spacing, factory personnel interviewed about work stations, or a medical staff asked to comment on the doctor's lounge. An intelligently prepared questionnaire in any of these fields will obviously not be limited to a single point. It will attempt to explore all of the aspects of the new environment that affect the users.

The advantages of such surveys are substantial. They are one obvious means of permitting people to participate in the decisions that affect their lives. They also afford the designer a means of seeing an entirely new set of relationships and values. It is like opening a door into a new world, the world of the user. It is not uncommon, however, to find that administrators and executives, the designers formal clients, exhibit some concern about user surveys. They fear that the survey will generate a host of extravagant demands that they will be incapable of satisfying. I can't say that this will never happen, but it has been our experience that when the users are asked to comment, their observations are eminently practical and matter of fact.

User surveys are by no means a complete answer to all the problems a designer must deal with. Users, for one thing, may not be acquainted with all the options that are available. They could hardly express a preference for continental seating in an auditorium unless they knew what continental seating was. In fact, their answers would be largely a matter of conjecture unless they had experienced such seating and were aware of both its advantages and limitations. A still more serious limitation arises from the fact that such surveys inevitably imply some pre-definition of the problem. In other words, if a new school or a new hospital is contemplated, the survey tends, by its use of such words, to focus attention on the formal systems that are traditionally implied by those terms and to foreclose consideration of radically different alternatives. Used alone, the user survey is not likely to stimulate creativity or to raise the question of whether learning and health care are necessarily confined to schools and hospitals.

There is yet another use of interviews and questionnaires; the measurement of such abstract factors as motivations, attitudes, images, goals, and strains. I must warn you that in introducing such factors into the planning equation we are drifting far off the beaten path. There is a considerable body of opinion that for the physical planner to address himself to such evanescent abstractions is an exercise in futility, a witless pursuit of the Holy Grail. While conceding that it is a highly uncertain undertaking, it is my own view that we must deal with such factors if we hope to break through the stereotype solutions that abound in architecture.

Much of this argument can be defused by turning from the abstract to the concrete. It should be clear enough that if we limit our questions to the restricted confines of a specific building type, we are channeling the responses. If, on the other hand, we invite the users to discuss the goals or objectives they have set for themselves in a given setting, and the strains they experience in achieving those goals, then we have opened the possibility that we can define a new kind of setting that will better assist these people to accomplish these goals. It sounds somewhat clumsy to carry on this discussion in terms of "settings" rather than "buildings," but it is done to stress a point. It permits us to keep an open mind toward the possibility that the result of this definition may be a building unlike any other or may not even be a building at all. This last eventuality is one that produces a sinking feeling in an architect's stomach.

If this concept remains fuzzy, I believe it will clear up when we consider the question of translating abstract data into design solutions. For the moment, however, we are still concerned with the methods of collaboration rather than the results.

If you were to combine in one survey demographic data on age distribution, family size and income, add to that a set of broadly defined facility descriptions, and conclude with a discussion of values, goals, and images, you would produce the information necessary to construct what behavioral scientists term a social-psychological profile. Since almost any project involves not one but several populations, administrators, teachers, pupils, and parents, or executives, supervisors, clerical staff, and custodians, you should actually produce profiles for each population. You would also have on your hands a dismaying amount of complicated data. At the present time, this use of social-psychological profiles appears to be the most sophisticated application of behavior research now being employed in the planning field. While further refinement and greater sophistication is possible, there is not much point in elaboration until the design disciplines begin to apply the procedures that are already available to them.

The potential avenues of collaboration that have been discussed, consultation, literature searches, observations, user surveys, and social-psychological profiles are all extremely valuable to the working designer. Each moves one step closer to an exact definition of a problem, a goal that seems to move farther away the closer you get. Any one of these techniques, however, is useful in itself. It would be extremely unfortunate if I created the impression that simpler measures were not worthwhile. The concerned designer will get all the valid information he can, but if he can't get it all he will take what's available.

Putting Theory to Work

Returning to the behavioral program for the new student center at Cal State, our first step was the design of the research procedure. The area of investigation had to be defined and the various populations to be studied had to be identified. This called for some policy decisions at the very outset. Although the project was wholly financed by the students and would, in actual practice, be used primarily by students, it represented a major addition to the campus and would in some degree affect the entire campus community. This suggested that the populations to be studied had to include, at a minimum, the administrators, the faculty, and the staff as well as the students. In addition to these, we included the alumni and the larger community in which the school was located. We felt that the campus was not an independent microcosm but a part of society. The school derived its support and its students from that society and its general welfare and the value of the degrees it offered would be affected by its standing in that society.

The area of activity observation was defined as all those parts of the campus that were not purely academic, administrative, or custodial. In addition we excluded the athletic facilities since their use was controlled and not spontaneous. This still gave us a considerable area to cover: all the outdoor spaces, the food service areas, and the lounges and leisure time facilities. By consulting the campus calendar, we were able to establish a schedule of observations that would cover both normal activities and special events within a daily range from 7:30 in the morning to 10:00 at night and a calendar range of three months. During this time observations were made on twelve different days.

The scope of the observations may seem somewhat ambitious in view of the traditional concept of a campus union as primarily a leisure time center, but it was just this traditional concept that we were trying to avoid. In general terms, we were attempting to identify all the non-academic, non-administrative behavior on the campus in order to produce a building that would accommodate that behavior. Such a building might be entirely unlike the traditional stereotype of a campus union but it would be a specific response for the needs of this particular university.

In conducting the observations, the observer followed a set course through the open spaces on the campus and through the interior spaces that were included in the study. In each instance he would note what, if anything, was taking place, who was involved, and what they were doing. At the same time he would photograph the settings. If he entered the cafeteria at 10:30 in the morning it would not be sufficient to note that there were seventy-nine people there. That might imply that there were seventy-nine people eating a late breakfast, when in fact they were not only eating but sleeping, studying, talking, playing cards, and putting on impromptu concerts. Unless such minutiae of behavior are recorded, it is impossible to get an accurate picture of what is going on. If repeated observations indicated that the cafeteria was consistently used in this variety of ways, then it would be reasonable to assume that it was serving as an informal social center as well as a place to eat.

In a similar sense, if the observer entered the courtyard of a building and found no one there it would not prove much. If repeated observatiions produced the same results, however, he could reasonably harbor the suspicion that the courtyard did not serve any useful purpose in campus life. This would not be much of a discovery, on the surface, but when you added to it the information about the spaces that were used you would begin to construct a description of the characteristics that make such places useful. That is a long step toward intelligent planning.

The use of photographs in our observation program deserves some discussion. Over a period of time it is easy to assemble hundreds of pictures of an extensive setting. In order to make our analysis as objective and efficient as possible we use simple thirty-five millimeter equipment, black and white film, and have each roll printed on double size proof sheets to provide an image big enough to study without magnification. It is surprising how much information can be obtained from a study of such proof sheets, even by someone who has never been near the site. Having the photographic data independently analyzed serves as an excellent check on the observer. An equally important function of the photographs is to transmit the study recommendations to the client. Photographs are much more persuasive than words in demonstrating the realities of human behavior, particularly when this behavior is at odds with our expectations. In such cases it is sometimes difficult to persuade clients that the architect is not hallucinating. The illustrative photographs, enlarged and mounted for display, are remarkably convincing.

The other arm of our Cal State study, the interview program, was under the direction of Tom Lasswell. The whole complicated procedure of selecting samples, preparing interview schedules, training interviewers, devising computer programs, and analyzing the results, was in his hands. The interview instrument that was ultimately developed consisted of a sixteen page questionnaire to be used by a trained interviewer in each interview session. The first two pages, covering demographic data, were designed for students only. The main body of the questionnaire covered institutional goals, student motivations in attending the school, values, images, perceived strains, an evaluation of existing buildings on campus, and an inquiry into desirable features that might be incorporated in the new building. Each block of questions listed a series of items to be evaluated or ranked but it began with an open-ended question to encourage the respondent to express his own feelings and opinions.

While I am not familiar with all the mysteries of sample selection, the student sample was selected by applying a table of random numbers to the official enrollment list. The administrative and staff samples were, in contrast, determined by the individual's position and function. Selecting a sample from the community at large was another matter. Los Angeles County contains over eight million people. To attempt any kind of extensive interview program with a statistical sample drawn from that enormous number would have consumed all our resources and would have delayed the completion of the study for months. The device that was finally employed was to select a group of community influentials in three localities of the region that had been identified as exhibiting different ranges of incomes, occupations, and social characteristics. The description of the key community influentials, ranging from labor leaders to the local newspaper publisher was modeled on a program devised for an earlier study by Dr. Samuel Stouffer. The three localities were drawn from a study of the Los Angeles area prepared by Dr. Marchia Meeker. When all the interviews were completed, the results were tabulated and processed by a computer. The physical product was an avalanche of multi-fold computer printout.

Making Architectural Sense Out of the Data

As a result of our field work we had a comprehensive set of notes and a sheaf of photographic proof sheets covering our observations, and a stack of multi-fold printout covered with the bizarre gibberish of the computer. In this form our information was of little value. It had to be analyzed for content before it could be put to any practical use.

Dr. Lasswell's analysis consisted of a report in which the responses of each population to each category of question was summarized. These were then evaluated in narrative form to establish the issues where very clear opinions emerged. This led to a profile or social-psychological model of each population. These profiles were then compared to determine their points of agreement and their points of conflict. The points of agreement produced a consensual model or a super-profile of the total population studied. Models that emphasized each populations' deviation from the consensus were also described. As a last step he compared conservative models with possible radical models. These terms have nothing to do with political philosophy. As they are used in this context, "conservative model" refers to the body of opinion that is satisfied with the existing situation and "radical model" is applied to the body of opinion that would like to see things change. With regard to the existing campus architecture, for example, the students produced a conservative profile (they liked the existing style of building) and the administration produced a radical profile (they wanted to change the style and appearance of the campus). This would appear to be something of a role-reversal but it probably illustrates how easily we can be misled by conventional wisdom. In the terms of this study, the student profiles frequently generated a conservative model and the administration profiles frequently generated a radical one.

Our own analysis of the observation data followed a somewhat different course than Dr. Lasswell's. Because this kind of material is not amenable to statistical evaluation our analytical activity can best be described as searching for patterns. The notes and photographs of each setting were studied to see what information could be derived from them. As this information built up it became apparent that certain kinds of behavior were recurrent; patterns of behavior began to emerge. Ultimately, it became possible to make simple declarative statements about the observations with some degree of accuracy. It is not, however, an automatic process; some degree of personal judgment is always involved.

The search for information in the observation data evolved in several ways. If it consistently appeared that students approaching a doorway loaded with books and briefcases had a hard time opening the door, the statement implied is very simple; "manually operated doors are inconvenient for students". This is a straightforward cue for the architect to supply automatic doors or eliminate doors altogether. Other patterns were a great deal more complicated. If two outdoor eating areas were consistently vacant and another one was always crowded, there had to be some difference between the areas to account for the disparity. The only way to solve the puzzle was to search for the difference. In this instance the popular area was the only one where food was served directly into the area. Taken together with other evidence, this led to the statement that "CSULA students usually consume food as close to the point of supply as possible."

The example of the unused courtyards already mentioned was part of the key to another statement. Our problem was to identify the differences between the empty courts and those that were teeming with people. After checking the facilities in each of these locations only one difference emerged; the courts that opened to the main campus thoroughfares were busy and those that didn't were empty. Statement: "Campus spaces that are not on the main thoroughfares are of little value to the campus community." Incidentally, in the process of arriving at that statement the main thoroughfares had to be identified first, a task which is not nearly as simple as it sounds. You certainly can't tell where they are by measuring the width of the walkways.

This kind of analysis continues until the information is exhausted or the observer is exhausted, whichever happens first. The observation process is capable of producing a bewildering amount of information. From our Cal State studies we produced statements covering campus communication networks, study habits, eating habits, social behavior, traffic patterns, ethnic territories, campus neighborhoods, personal contacts, and the curious process by which students allocate their free time.

The information contained in Dr. Lasswell's report had to pass through a somewhat similar process. A series of statements was derived from his analysis to form a set of social-psychological criteria derived from the interview data. Again, some of the statements were fairly obvious while others required protracted study. It was no real surprise that of the various strains and pressures afflicting the students lack of money ranked high. It obviously led to a statement that "Low-cost and no-cost activities and services should receive the highest priorities in planning for students". This apparently innocuous declaration assumed considerable stature during the actual planning process as it served to sort out what was to be included and what left out.

The students reported other strains that were not as easy to deal with. At the very top of their list was the issue of "Freedom versus responsibility," a source of stress that any adult will recognize. They also showed a surprising bitterness about vocal students receiving more than a proportionate share of attention. Examples like these illustrate how quickly social-psychological data shift from rather obvious design problems that can be resolved with common-sense statements to problems that are extremely complex. The question arises whether or not an architect can actually deal with such problems. Certainly there is nothing in his conventional stock of design devices that will in any way alter a dominant or aggressive personality or relieve the student of the responsibility he assumes by becoming a student. If an architect can respond at all, it will be within the limited confines of a specific project in which he can exert some control. He can arrange the spaces he controls so that it will be difficult for anyone, vocal or not, to dominate the setting. He can insure the maximum number of personal behavior options or choices so that the student can exercise his preferences without coercion. The statements on these two points take the form: "Public spaces should be arranged so that they cannot be easily dominated by an individual or group" and "The total setting should offer the maximum range of behavior options from seclusion to group activities." These may seem like very minor responses to major problems, as indeed they are, but they are certainly better than no response at all. Within the confines of this project, at least, students will be free of some of the annoyances they deeply resent.

By patiently analyzing each finding in the report in this manner, a set of behavior statements was developed that began to imply some surprisingly precise planning guidelines. The final step in preparing the behavioral pattern was to describe the facilities that were implied by the statements. This was not done in the dimensional manner that is customary in building programs. No room sizes or floor areas were given. The description simply stated the kind of activities that should be accommodated and the nature of the facilities that would satisfy these needs. Again, the material ranged from the direct and obvious to the obscure and complex.

Based on the observation data it was obvious that the new building had to be located on the main campus thoroughfare if it was to serve the most people in the most useful way. That was a very important conclusion which led to the use of a site that suffered from severe limitations in other ways but was immediately adjacent to the main stream of traffic. While the data thus clarified the location, it was not equally helpful in demonstrating that a single structure should be the end product. There were certain kinds of services used by the entire campus community that might best be distributed along the main walk in a series of shops or kiosks. There were some short-term leisure time activities, including food and beverage service, that would serve the demonstrated need best if they were dispersed in the informal sub-campuses that had spontaneously sprung up around certain centers of student concentration. Yet there was a strongly expressed need for a focal point of non-academic activities, a center that would satisfy this particular student body's deeply felt concern that in the intensity of their drive for career training they were missing the social expansion and the intellectual enrichment that are the implied promise of higher education.

This set of factors presented a real dilemma. The study had demonstrated that while the students might express certain preferences or choices, their actions did not always support their expressions. They might express a desire to participate in more cultural activities, but the evidence indicated that the only things that could be counted on to influence their actions were the goods and services that were essential to their college existence or satisfied their personal needs. As a result, if we had stripped out the service elements, the bookstore, post office, food service, information center, and others, and distributed these around the campus, the remaining facilities would have seen very limited use. Our decision to combine them all in one complex was based on the conclusion that the expressed need for a social, intellectual and cultural focus was a real one. In one sense, we needed the services at the same location to act as bait, to draw the maximum numbers of users who might then participate in the other activities of the center. This decision may appear questionable to some. It is an obvious use of the planning process to manipulate human behavior. There is one vital difference, however, between this effort and the more usual kind of manipulation that occurs in buildings; in this instance the users had defined the objectives. Our role was to develop the configuration that would best serve their interests and concerns.

* * * * * * * * *

The described process for developing a social-psychological program may seem exceptionally long, complicated, and expensive. It is all three. The real question is whether the results justify such an elaborate addition to the planning process. In my opinion the results surpassed our fondest hopes and I believe that everyone involved in the study, from the administration to the student representatives, would agree.

While it is hazardous to discuss costs in fixed terms during this time of inflation, relative figures have little meaning. The study I have just outlined, the observations, interviews, data processing, analysis and summary, cost $15,000. Compared to a project budget of more than five million dollars, it was a modest amount. It cost considerably less than the hardware on the doors, or the paint on the ceilings. It would not have to be extremely productive to be worth more than these items in terms of human values.

There is, unfortunately, no way to measure the benefits that will accrue to the campus in money terms. The results will be measured in terms of increased social contact, a reduction in stress, and the development of new attitudes about the institution. What their value might be is a matter of conjecture, but anyone who feels that a building program should reflect, as accurately as it can, the aspirations and needs of the users will rank them highly.

Perhaps the value of behavioral programming can best be understood in contrast to the procedures that would normally have been followed. Without such a study the design of the building would have been based on a survey of architectural solutions to similar problems and whatever intuitive judgments of need might have been made by the planning committee and the architect. Granting these individuals all the good-will, motivation, sensitivity and talent in the world, they suffer the same limitations of experience and mind-set that are common to all mankind, It is only through such comprehensive programming process that the faceless, unknown users can be identified and elevated to the position of a reference group with needs and opinions that are stated with as much clarity, if not with the same personal force, as is employed by the people who are actually involved in making the decisions.

For the architect, the drawbacks of this lengthy process are significant but the benefits are great. His work is complicated by a set of strange new factors, but he is no longer forced to rely solely on his own intuition, or on the conventional wisdom of the design professions for guidance. Anyone who sees architecture in terms of human beings rather than as physical artifacts will recognize the benefits.

Chapter V

Tracking Down Solutions
(And Side-Stepping Stereotypes)

Hooper Avenue School is an overaged elementary school in southeast Los Angeles. In the terminology of the day, the neighborhood would be called a black ghetto though I would prefer to refer to it as a segregated, low-income neighborhood. After months of close contact with these people it would be personally embarrassing to apply a label to them that might in any sense be considered derogatory. In the course of developing a plan for rebuilding this school we had conducted a social-psychological profile study, again in collaboration with Dr. Thomas Lasswell, and had developed a set of social-psychological criteria as the basis for our building program. As is the usual case, some of these were rather simple and obvious while others raised very complicated issues.

In interviewing the adults in the neighborhood some very strong preferences were expressed for programs and facilities that the school was already providing. This seemed to indicate that the neighbors didn't really know what was going on at the school. The same interviews indicated that the community's gravest concern about the school was the control of vandalism. In this they were not alone. Officials of the school system were doing everything in their power to control vandalism and their efforts were probably matched in every urban school system in the country.

The criteria implied by the above facts were clear enough: improve communications between school and community and control vandalism. What an architect might be expected to do about them was another matter. We could turn the schoolgrounds into a fortified

Public sidewalk through the school:

A model of the new buildings at Hooper Avenue School showing the public sidewalk across the schoolground (arrow) and the little park areas provided for the community. Buildings along the street frontage are existing. The gazebo-like structure where the walks intersect is an on-site field trip center, a place where mobile displays can be brought to the school. This is our response to the high educational rating given to field trips by the community.

encampment in the hope of reducing vandalism, but the defensive measures that had already been taken didn't seem to work very well. Yet this same community had very little trouble with malicious mischief in the commercial district only a block away. We could urge the school administration to open up better channels of communication with the neighborhood, but this would be an operative response rather than a planning response.

The ultimate result of these concerns about the problems of communication and vandalism at Hooper Avenue School was an extension of the public sidewalk across the schoolground. Other studies we had made of public behavior indicated that the surest way to generate pedestrian travel was to provide a short-cut for through traffic. A short-cut across the schoolground would be a convenience to the neighbors and would give them a much better idea of what was actually going on at the school. At the same time it would give the adult community an opportunity to exercise some control over vandalism if they were as seriously concerned about it as they appeared to be. If community pressure was capable of controlling vandalism on Central Avenue, it might accomplish the same result at the school if it had a chance.

This example illustrates the nature of the problem the architect faces in translating the abstract data of social-psychological research into three-dimensional physical solutions. Neither of the criteria outlined above suggests a familiar response with an assured result. Conventional design devices do not offer ready solutions to these problems. In proposing a public sidewalk as an answer we were advancing a carefully considered hypothesis that seemed to fit the facts at hand, but only a test of this hypothesis in use could determine its validity. If it worked, we would have made one more bit of information available to the designers of urban schools with similar problems. If it didn't work we would have to close the gates and would be right back where we started. When you are experimenting on public projects it is a good idea to work out a fail-safe system.

In effecting such translations the designer frequently finds himself in a similar situation. He has available to him a host of conventional solutions to apply to problems as they are usually defined. When these same problems are redefined through the process of social-psychological research, he is apt to find that the conventional solutions no longer fit. As a result, he is faced with the problem of originating new answers, a process that is by no means clear and precise and that will surely test his creativity and the scope of his knowledge about human beings. There are some techniques that will help to organize his thinking, but not many.

Translating Words Into Forms

In Tom Lasswell's summary of his interview study at Cal State is one paragraph that pinpoints the nature of the translation problem:

> One of the most striking findings of this study is the consensus with which all samples conceptualize the primary goals and student motivations of CSULA. It is seen as a future-oriented training agency, with only the faculty adding the dimension of an independently viable community of thinkers and scholars providing a special intellectual climate for persons who wish to explore various fields of knowledge. The policy decision that may be indicated here is whether to pursue the conservative consensual model, the conservative faculty model, or a radical model.

At that point the architect is still a long, long way from the drawing board. The last sentence offers three strikingly different options. After determining what each of these models may be and ferreting out what they might mean in physical planning terms the even larger question of who should make such decisions remains.

In view of the complexities inherent in bridging the chasm between the abstractions of values, goals, and strains and the concrete realities of construction, Robert Sommer has suggested that a new breed of specialist, a professional translator with one foot in both camps, will ultimately emerge to fill this need. That may well be the future course of events, though the prospect doesn't fill me with enthusiasm. Injecting one more layer of interpretation into the process simply moves the designer one step further from the people whose needs he is supposed to serve. It is my hope that the behavioral scientist-designer team can develop the necessary skills to do this job themselves. There is a sense of immediacy and reality in dealing with the original data and participating in the analysis of it that is highly stimulating. It could easily be lost if the data were passed through an intermediary.

One of the steps a designer can take in dealing with the masses of data generated by a major study is to sidestep the global questions about which model to pursue and deal with the information in smaller chunks. At least this is the process that we have employed, and it has been our experience that if the items are all considered one by one or in small sets, by the time the process is completed some of the larger questions will have resolved themselves.

One kind of sorting that is useful is to group information around a *behavior setting*. In our Cal State study one of the most consistent behavior patterns we observed was the student use of outdoor spaces for study. This was not true of just any outdoor space, it was highly focused at the entrances to major buildings. We also observed that group study was a common practice and that where it was possible to do so, many students ate snacks or consumed beverages while studying. On the basis of the interview data that revealed the serious time pressures on these students due to off-campus work and commuting, we concluded that this practice of studying wherever they happened to be—sitting, standing, or leaning—was a more efficient way of using small segments of free time than going to the library. The reasons, however, are unimportant. What is important, empirically, is that it occurred.

Taking the entrance or forecourt of the student center as a behavior setting, it was possible to make several decisions about the design characteristics it should have as a direct result of this observed studying behavior. Since the student center was clearly a major destination it could be assumed that the entrance area would be used for study.

In that case, some provision for convenient and comfortable study should be made; there should be comfortable benches and the benches should be provided with tables for spreading out books and papers and tablet arms for writing on. Since eating was so customary a part of study, there should be outdoor food service in the area. You may recall an earlier discussion of the way straight benches inhibit group formation. This factor is just as decisive on a college campus as it is in a public park. As a result we could stipulate that the seats and benches in the entry plaza should be arranged in facing pairs, and re-entrant angles to permit easy group study.

Bit by bit our process built a description of the physical facilities that were necessary to accommodate one customary type of behavior, outdoor study, To expand our description of the requirements of this entrance area we could add the information that most students harbored a deep resentment against vocal students commanding more than their share of attention, leading to the conclusion that the plaza must be broken up into small areas that were not easy to dominate. Plug in the preference for shade versus sunshine, and student hopes for more informal contact with the faculty, and a space is defined that is shaded rather than open, with an entrance that focused traffic rather than dispersed it. By the time all of the factors that apply to this behavior setting were assembled, the features of the design were largely established. The entrance court, or plaza, or whatever you might wish to call it was no longer just a formal appendage to the building. It had been defined as a behavior setting with a clearly defined set of functions to fill. While there were still decisions to be made regarding color, material and form, the nature of the setting and the way it was to function had been largely settled.

Another useful form of data sorting is to group information around *behavior circuits.* A behavior circuit can be defined roughly as the series of actions that are involved in carrying out a personal mission. A full circuit for someone going to the bank, for example, would start with assembling the necessary information, passbooks, and checks before leaving the house or office, going to the bank, transacting his business, returning to the home or office, and putting away his banking materials. In designing a bank, of course, attention is centered on that portion of the circuit that takes place in banking quarters. Based on observations of the typical behavior of people entering a bank the conclusion can be drawn that they will normally scan the teller stations to see if there is an open window and teller free to handle their transaction. This suggests that the teller line should be positioned so that all the windows can easily be scanned from the entrance. If a teller is free, then it should be possible to move from the entrance directly to the open station. If all the stations are busy, then the customers will get the fairest treatment if they move into a master queue so they can be dispatched to the first free station.

This set of conditions dictates an arrangement with special characteristics. It suggests the design of some kind of contraption, a piece of furniture perhaps, that can be used to form a queue during rush hours and yet be wide open during periods of light traffic. It is possible to define this contraption with considerably more precision as other factors are added. Many of the people who enter the bank will be burdened with parcels, handbags, or briefcases so that it would be a convenience for them if there were a place to rest these parcels while they are waiting in line. Many of them will fill out some form or deposit slip before proceeding to the window. It would be an added convenience if they could do this while they are standing in line. Since they will be in a moving line, this implies some form of continuous surface to write or work on as the line shifts. Given the general tendency for people to lean on anything at hand while waiting, we can add a further stipulation that this new construct be sturdy enough so that it won't collapse under the load.

A Bank Lobby Queuing Device:

When lobby traffic is light the wings swing back under the check stands and customers can walk directly to the teller windows. When the wings swing out they form a master queue. Filling out deposit slips and endorsing checks can be done in the line.

Add to this the standard requirements for pens, forms, calendars, waste receptacles, and ash trays, and you have defined a new piece of furniture or equipment that will accommodate the observed behavior of customers entering a bank. You won't find anything quite like it in a furniture catalogue. It is a specific artifact designed to accommodate a behavior circuit that is normal in Southern California banks. I have no idea whether this circuit occurs anywhere else in the country. I suspect it does but that really isn't important. By identifying such behavior circuits, it is possible to describe the physical facilities that will make it easier for people to carry out their personal missions.

Behavior settings and behavior circuits each provide a convenient focus for organizing information that aids the translation process. There are other types of data generated in social-psychological research, however, that do not fall in either category. These are the abstract questions of goals, values, motivations, and images. Translating these into physical solutions is frequently difficult and sometimes impossible. Even when an answer is possible, it often lies outside the scope of the project or is beyond the designers's control. However, the thought processes that are followed in making the translation are exactly the same as those used in analyzing more objective data: definition of the problem and a search for a fitting response.

A relevant example is supplied by our observations at Cal State. The photographic evidence demonstrated that the graphic communication systems on the campus were not only numerous, but surprisingly extensive. Almost any setting adjacent to the main thoroughfare would exhibit one or more of these communication channels: newspapers, bulletin boards, poster kiosks, and graffiti. In addition, the campus would from time to time sprout card tables, like mushrooms that spring up overnight, when an organization felt the need to communicate its program or campaign in a more personal or persuasive way than could be done with bulletins or posters. It was fairly easy to define these formal channels of communication; the newspapers were a channel from the world to the individual, the posters and card tables provided media for groups or organizations to address individuals, and the bulletin boards were avenues for contact between individuals. That left one glaring void in the network. There was no formal channel available to the individual who wanted to address the world at large.

This need was filled in an informal way by the graffiti on the construction barricades. A close study of the content of these scrawled messages revealed that they had some consistent characteristics. They were, first of all, confined to the areas of heavy traffic. No one bothered to write on the barricades unless he could expect a good audience. They were also confined to temporary surfaces. In

a.   Barricade debate. Open to everyone. It reads (roughly) from top left to right, down, and concludes (for the moment) at bottom left with a barely discernible expression of sympathy for the wife.

b.   Barricade art. This kind of light-hearted communication may not carry a profound message but its bound to make you feel better.

Barricade Displays. A stranger, or a new student, could tell more about the mood of the Cal State campus and the diverse viewpoints it espoused from the graffiti on the construction barricades than from any other source. We are incorporating a series of renewable graffiti panels along the main campus walk so the individual will have some place to tell the world whats on his mind.

some mysterious way, the campus respected an unwritten law that it was not appropriate to write on buildings. Last, the messages were coherent, timely, and there were no real obscenities. While there were examples of the usual political slogans and apocalyptic pronouncements about the fate of the world, for the most part the commentary was germane, witty, and often extended to lengthy written debates. Any newcomer to the campus could find out more about its preoccupations and the wide range of opinions it espoused by looking at the graffiti than from any other single source.

Once the nature of the CSULA's public communication systems

was described it was not particularly difficult to translate the information into three dimensional designs. Providing places for news vending, bulletin boards, poster displays, even for the ubiquitous card tables, was simple enough. Accommodating the graffiti, formalizing the informal, called for reachable surfaces that were either temporary or renewable and located adjacent to a concentration of traffic. The problem in this instance was not in translating the data but in recognizing that the graffiti was a legitimate form of communication that deserved recognition. The resulting panels, positioned alongside the main campus walk, provide an outlet for individual expression that has never had an outlet before.

Another form of abstraction encountered by environmental designers are the goals that individuals and organizations set for themselves. These are often expressed in terms that are so abstract as to afford no guidance at all. "To make family life more meaningful," or "To form integrated personalities" represent a class of global generalities that can be counted on to drive an architect right up the wall. If such phrases are to have any meaning at all they must be broken down into discrete statements bearing on human behavior before any translation is possible. This is sometimes very hard to do. One of the reasons it is difficult for some people to articulate their goals in a usable fashion is that they are ill-defined in their own minds. When an architect encounters that situation he faces a long, laborious process of goal interpretation.

The field of day care for young children is one that is currently receiving a great deal of attention. The mission of child care and day care centers has been extensively probed, studied, analyzed, and debated. Yet, when the time comes to design such a center, it is very difficult to find precisely stated objectives. A lot of advice is available about centers that have been built and there is an ample supply of solemn generalizations about "helping the child through the socialization phase." Such a statement may be perfectly lucid to someone who has made a career of caring for young children. To an architect it is as impenetrable as Runic script. It requires a lot of decoding, or breaking down into its basic components, before translation is possible.

Cracking this code is a lengthy process. For purposes of illustration it is sufficient to deal with just a few aspects of the socialization process. cooperation, responsibility, a willingness to help younger children, and expansion of skills and confidence. Let's add to the mix the stipulation that socialization is a joint operation involving both parents and teachers who, of necessity, must have some way of keeping in touch with each other. Broken down even in this crude form the original goals become much more manageable. If these small children are supposed to learn to cooperate, then it might be useful to have certain aspects of the care center designed so that they are only operable by a cooperating pair. The callous mind will immediately advance the proposition that the door to the playground should be designed so that it requires two or more children to open it; cooperate or don't play. This is unnecessarily coercive, however, and apt to be traumatic for some youngster who happens to be, at the moment, without friends. The same principle could be applied with less stress by stacking the cots or the nap pads in an enclosure that required two children to get them out.

It will be easier for the child to learn responsibility for his own things if he has a separate space of his own to keep them in. He may be induced to help younger children if helping gives him a chance to demonstrate an advanced skill, like priming and operating the hand pump at the two year old's sand box. Challenging settings to test and expand his motor skills and confidence can be arranged by establishing a desirable goal, a special book or game nook, that can be attained by either a conventional set of steps or an unconventional rope ladder.

Bringing the parents into contact with the staff is not hard to do. The entrance can be arranged so that it is only negotiable by an adult. This will force the parent to escort the child to the front door to operate a high gate latch or open the door itself. If someone from the staff is on hand at that point, a useful interchange of information can take place. Such a solution would, however, ignore the problems of the parent. Early morning is a difficult rush period for many parents. Forcing them to detour to the front door at a time when they may already be late for work would be extremely coercive. It is just the kind of configuration that leads to the hanging of architects in effigy. An alternate arrangement that requires the parent to enter the building to pick up the child in the evening, when time pressures are not so severe would meet with less resistance.

Breaking down goals and objectives into behavioral program statements makes it possible to design buildings that are highly supportive. It is true that child care specialists have techniques for dealing with any of the points cited above without relying on an assist from the building. Yet, when the physical spaces can be so easily arranged to make a positive contribution to these goals, it would seem to be a foolish waste of resources to ignore that potential.

It is inevitable that human studies dealing with abstractions will define some goals that lie largely or completely outside the architect's jurisdiction. That doesn't mean that they are unanswerable but that the answer has to do with the way the system functions or the way the building is operated rather than the way the building is arranged.

The desire for greater recognition of black accomplishments and black culture in the school curriculum that was expressed by the

Hooper Avenue neighborhood falls in that category. The focus on future goals, the career-orientation that motivated the Cal State student body, is another instance in which the physical facilities were only related to the goal in a tenuous way. That doesn't mean that these aspirations should be ignored. While this information is most useful in the direction and administration of a facility rather than the design, so long as there is any possibility that the design can support these goals, even in a modest way, the designer has an obligation to contribute whatever he can.

The Cal State study provides an illustration of how persistence pays off in translating this kind of data. There is very little obvious relationship between student's interest in career training and the design of a new campus union. It does suggest that in programming events for the union, it would be useful to bring representatives of a wide variety of career opportunities to the campus in a context that would permit the students to have the greatest degree of personal contact. Operating a lecture program is not a new activity for this type of building. Many campus unions make an effort to bring outstanding speakers to their campus. The only variation that a union direction might derive from our evidence is that the normal diet of personalities and political figures might be varied to include speakers representing a wide variety of professions and occupations.

A normal design response to this need would be to incorporate in the plans a series of meeting rooms of various sizes to accommodate lectures and informal discussions. It is possible, however, to go somewhat further than this. Segregated meeting rooms housing scheduled events tend to attract an audience that already has some interest in the subject. They may not serve to expose the maximum number of people to new ideas and new contacts. If we wanted to offer the student body the widest exposure to career options, then the seminars and lectures should be scheduled at a point that carried a heavy volume of student traffic.

That conclusion is not wholly without foundation. Our observations had demonstrated a persistent behavior pattern of students who were approaching public events on the campus. When these events were open, and did not require a ticket or paid admission, a large percentage of the students approach them in an uncommitted, tentative way. They would stand against the back wall, lean over the railings, or clog the entry. The best description of their actions would be that they were sampling or testing the performance before committing themselves. If they liked what they saw, they eventually drifted into the space and found a seat. If, in comparison to the other options that were available, the event did not justify the expenditure of their limited free time, they moved on. Any open event could expect to have this kind of floating participation.

A setting intended to accommodate floating audience behavior should have certain definite characteristics. It should be adjacent to a concentration of traffic for the greatest exposure. It should be approachable from several sides. It should accommodate a group of ten or a group of one hundred with equal ease. Of these factors, only the need to accommodate a flexible group offered a design problem. It could be done conventionally by moving chairs around but that would affect the spontaneity of the gathering.

Our answer was derived from yet another aspect of student behavior, the tendency of students to arrange themselves vertically as well as horizontally in order to make a compact group. In their spontaneous social formations they sit on the ground, the steps, the handrails, rubbish containers, seat backs, or anything else available in contrast to the more normal horizontal deployment of their elders. As a consequence, an expansive setting in student terms can be accomplished

An "Upholstered Bleacher":

Carpet covered, to permit sitting, standing, or sprawling at a number of different levels this sectional seminar circle is on dollies so that it can be located where the traffic is concentrated. It makes it possible for anyone passing by to sample the action before deciding whether to join in.

by adding more vertical layers rather than expanding in size. Based on this clue we designed multi-level lounge seating, covered with carpet, that accommodates every conceivable student posture. Like the queuing contraption developed for the bank lobby, this piece of furniture has no name, though it has been ungraciously described as an upholstered bleacher. Whatever the name, it satisfied all the criteria: it can be positioned in a lounge, a concourse, or mall, it is approachable from all sides so that standees can participate, the segments can be arranged in a circle, a semi-circle, or in facing quadrants, and it provides five levels of seating.

I will concede that the design of this special furniture has very little to do with the career concerns of the students at Cal State. It can do nothing directly to raise their future salaries or to find them better jobs. All it can hope to do is to accommodate their behavior patterns and make it possible for them to contact individuals and ideas that they might otherwise miss. Even that limited objective is dependent on operating policy. Unless someone makes the effort to invite the right kind of people to the campus for informal meetings with students our upholstered bleachers will be just odd looking lounge furniture. On the other hand, we have responded to a need that could never be served quite as well by any other means.

This example illustrates a fact that has recurred constantly in our work; in designing behavior settings it is impossible to separate the design or selection of the furniture, equipment and accessories from the design of the structural enclosure. It is not uncommon to find these two activities segregated in the minds of both architects and clients. The behavior setting concept cannot afford to make a distinction between the two. Interpretation of the social-psychological data frequently requires both a special configuration of the building and a special form of equipment to satisfy the need.

Getting the Problem to Describe the Solution

It is obvious from the preceding examples that the process of translating abstract social-psychological data into three dimensional solutions is anything but an automatic process. It will probably take an analysis by a cognitive psychologist to define the exact mental steps that occur during translation. In broad terms, it appears to consist of a search for fit, a kind of mental matchmaking. Given a specific human factor or set of factors, the translator conducts a search through his own mental archives for a construct or form that will accommodate it. His solution will obviously reflect the state of his archives and his ability to synthesize new forms from his stored data. Under such circumstances, it is entirely possible that an independent analyst working with the raw information that we have been discussing would come up with better solutions, alternate solutions, or no solutions at all.

In view of the uncertainty that surrounds translation, any procedure or system that would reduce the uncertainty would be a great boon. While nothing even approaching an automatic process is available, there are some techniques that are decidedly helpful. They deal with methods of organizing information in such a way that, while the problem does not solve itself, the designer is forced to recognize and deal with each of its elements. Furthermore, these techniques provide a graphic means of charting translation decisions, an activity which is in itself very helpful in synthesizing complex data.

The most comprehensive and systematic proposal for dealing with the translation problem is that developed by Christopher Alexander. His book *Notes on the Synthesis of Form* deals with the entire design process, but for our purposes we can focus on the part that describes the conversion of verbal statements of design objectives into three dimensional solutions. Essentially, the process consists of organizing design criteria into sets and identifying the links or relationships between elements in a set and between the sets themselves.

In accordance with Alexander's concept a full set of design criteria for a single family dwelling would include not only the social-psychological factors that we have been discussing but all the other elements that would have to be considered in the design; functional considerations of food storage and preparation, house-keeping and maintenance problems, waste disposal systems, community regulations on land use, and so on. These elements are then broken down into sub-sets that assemble all the related elements of the full set. One sub-set might contain all the elements that deal with motor vehicles, including access, deliveries, and parking the family car. Another might group the factors having to do with food storage, preparation, and service. A third might deal with children's play activities and a fourth might deal with privacy requirements. These sub-sets might encompass enough elements and enough diversity to be broken down into sub-sub-sets or minor sub-sets.

Once the sub-sets of design criteria are defined an analysis must be made of their relationships which may be either positive or negative. In some instances a positive link may be mandatory. Vehicular access to the house requires a positive link to the public road system. A positive link between the car parking area and the food storage area is also mandatory because most food arrives at the American home via automobile. The parents' bedroom and the children's play area, on the other hand, are negatively linked. So long as there is any possibility of nighttime working and day time sleeping it would be desirable to keep noisy areas at some distance from the bedroom.

Once the sub-sets and the link relationships between them have been established it is possible to display this information in graphic form. At this point, while it may not be in true scale, Alexander's graphic representation of the sets and links defines all the relationships that should exist in the final plan. In a sense it *is* the plan, and requires only the addition of boundaries, access points, and structural systems to complete the three dimensional concept.

This is a highly simplified description of a complicated process. Even in such a relatively simple project as a single family house there are a surprising number of criteria and links to deal with. In a larger project such as a whole neighborhood or a whole community, the amount of data to be organized and analyzed is truly formidable. One of the virtues of Christopher Alexander's method is that by systematically organizing the data in accordance with the principles of mathematical set theory at least part of the process can be transferred to the computer.

Alexander has subsequently added another concept to the process that promises to be of great benefit to the planning professions. One of the signal weaknesses of these professions is the lack of a method for organizing information developed for one project in a systematic fashion so that it can be transmitted for general use in similar projects. It may seem odd that these professions, which employ so many sophisticated techniques in the normal course of practice, have no coordinated data bank of information to draw on, but this is the case. Alexander's answer to this problem is the development that he calls "Pattern Language." As each item of the design criteria is analyzed and translated, a simple sketch of the resulting configuration and a brief summary statement of the criteria and the consequences are prepared. When all of the criteria have been subjected to this process the designer winds up with a volume of single page briefs that summarizes the results of the translation process. In any later project where similar criteria appear he has at hand a convenient and accessible source of information. What is even more important, the information is in a form that makes it available to anyone else who is wrestling with a similar problem. The benefits that would accrue to the design professions if this procedure were systematically followed and the information openly exchanged would be enormous. The benefits that would ultimately accrue to the public would be even greater.

Using Alexander's form of notation we can rather easily construct a pattern statement out of the information we have about the tendency of students to study outside.

In this instance, the sketch is not essential. There are a number of configurations that will achieve the same result. The designer of any similar structure has only to establish that he has a similar set of

IF: There is a major building entrance facing a primary campus traffic route,

THEN: Outside of each entrance an area will be set aside for student study with appro-
            priate furniture and equipment.

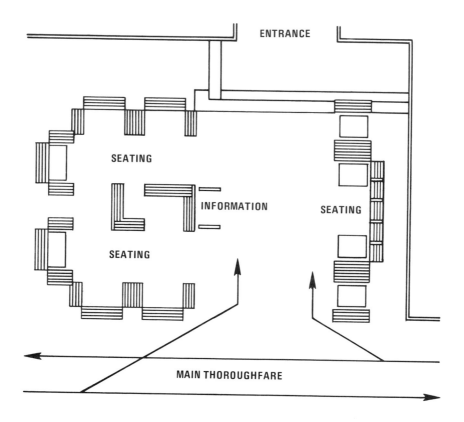

PROBLEM

One of the consistent behavior patterns identified in the study of the Cal State campus
was the tendency of students to study out-of-doors. While this activity was observed
in a number of places it was concentrated at the entrance to buildings that housed
activities that were essential to students as long as these entrances opened to the main
campus thoroughfare. The benches, seats, or other outdoor furniture provided should
have tablet arms or tables and should be arranged to permit either group or solitary
study.

behavior criteria to work with before applying this pattern in a way that is suitable to his project.

In our practice we have found it useful to modify this form of presentation somewhat. The proposal embodied in the pattern plan shown, synthesizes in one drawing a number of facts about student study habits on a specific campus. Whether our clients are suspicious by nature or we need to bolster our decisions by displaying our facts conspicuously, we have found it helpful to demonstrate graphically the steps that led to the development of the pattern. Each bit of information, such as the need for flat writing surfaces during study or the arrangements that make group study comfortable and efficient, is summarized in a single statement. Each statement is accompanied by a small sketch and photographs or other evidence that indicates the source. The final pattern is a synthesis of each of these statements. The end result is substantially the same as Alexander's notation but all the steps that were taken in arriving at that point are spelled out. This procedure is somewhat tedious but the results are very helpful to anyone who is checking our conclusions. Pattern development displays are particularly helpful to clients who need to feel that our recommendations have a rational base and are something more than hallucinations.

The value of information in pattern form may not be immediately apparent. This particular example seems so self-evident that it is hardly worth recording. Yet none of the buildings on the Cal State campus or any other campus I have ever seen provide accommodations for study in this detailed way.

Christopher Alexander's analysis of the design process is so revealing and his proposals are so useful that they should become a part of every designer's repertoire. Certainly they deserve a more comprehensive consideration than I have been able to give them here. Yet it should be pointed out that, useful as they are, they do not afford an automatic solution to the translation problem. At the point where written data is transformed into a three dimensional concept, when the problem becomes a solution, there is a gap that cannot be crossed by set theory, matrix analysis, or the computer. All these things may help, but the translation step is the product of a synthesizing mind. In the end, Christopher Alexander's design solutions are still the product of his own talents and the talents of his associates at the Center for Environmental Structure rather than the product of a process.

It may appear that a rigorously systematic procedure for stating criteria, sets, and links would complicate the designer's work considerably. If it does, it is only because he may have been accustomed to glossing over these requisites. The criteria, sets, and links are the same for any real life problem whether they are analyzed or not.

OUTSIDE STUDY AREAS

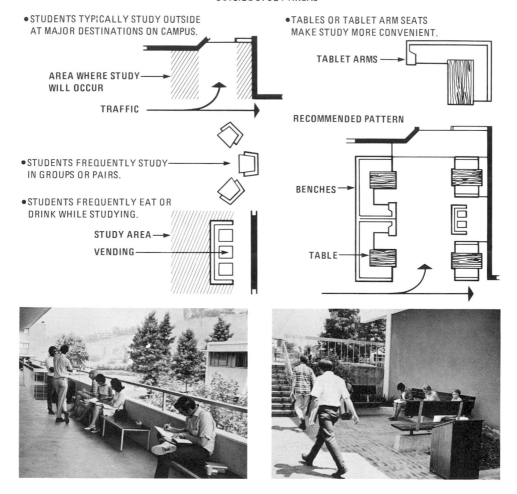

- STUDENTS TYPICALLY STUDY OUTSIDE AT MAJOR DESTINATIONS ON CAMPUS.

AREA WHERE STUDY WILL OCCUR →

TRAFFIC →

- STUDENTS FREQUENTLY STUDY IN GROUPS OR PAIRS.

- STUDENTS FREQUENTLY EAT OR DRINK WHILE STUDYING.

STUDY AREA →

VENDING →

- TABLES OR TABLET ARM SEATS MAKE STUDY MORE CONVENIENT.

TABLET ARMS →

RECOMMENDED PATTERN

BENCHES →

TABLE →

Translating research data into a building form:

a.   A pattern in the process of being developed. The pattern deals with the tendency of students to study outside on this particular campus (California State University at Los Angeles) and is based largely on the photographic evidence that is grouped around it. It has a clear bearing on the design of a building entrance.

INFORMAL STUDENT - FACULTY CONTACT

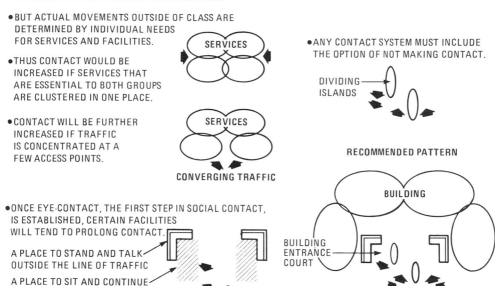

- BOTH FACULTY AND STUDENTS EXPRESS A NEED FOR MORE CONTACT OUTSIDE THE CLASSROOM.

- BUT ACTUAL MOVEMENTS OUTSIDE OF CLASS ARE DETERMINED BY INDIVIDUAL NEEDS FOR SERVICES AND FACILITIES.

- THUS CONTACT WOULD BE INCREASED IF SERVICES THAT ARE ESSENTIAL TO BOTH GROUPS ARE CLUSTERED IN ONE PLACE.

- CONTACT WILL BE FURTHER INCREASED IF TRAFFIC IS CONCENTRATED AT A FEW ACCESS POINTS.

- ONCE EYE-CONTACT, THE FIRST STEP IN SOCIAL CONTACT, IS ESTABLISHED, CERTAIN FACILITIES WILL TEND TO PROLONG CONTACT.

A PLACE TO STAND AND TALK OUTSIDE THE LINE OF TRAFFIC

A PLACE TO SIT AND CONTINUE THE CONVERSATION

- ANY CONTACT SYSTEM MUST INCLUDE THE OPTION OF NOT MAKING CONTACT.

SERVICES

CONVERGING TRAFFIC

DIVIDING ISLANDS

RECOMMENDED PATTERN

BUILDING

BUILDING ENTRANCE COURT

CONVERGING TRAFFIC

CONVERGING TRAFFIC

b. A second pattern statement has been added to the display (on the left). It deals with the desire of both students and faculty for more informal contact outside the classroom. This was derived from the interviews and has no supporting photographs but it also has a bearing on the design of a building entrance.

c. A model of the University Union at Cal State L.A. The entrance to the building reflects the two patterns illustrated above. The design of the entrance plaza is the result of additional considerations that are discussed in the text.

The following pattern displays, are part of a complete set developed for a new office building. Each pattern synthesizes the information from three sources, observation, interviews, and the literature of the behavioral sciences.

PERSONAL WORK STATION

- PROVIDE MAXIMUM WORKING SURFACE WITHIN ASSIGNED AREA WITH AS MUCH FLEXIBILITY FOR ARRANGEMENT AND ADAPTABILITY AS POSSIBLE.

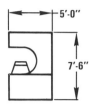

- LOCATION FOR OFFICE EQUIPMENT:
  TYPEWRITER   DESK CALCULATOR
  TELEPHONE   DICTATION UNIT

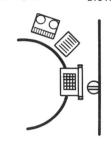

- PROVIDE LOCKABLE STORAGE FOR PERSONAL ITEMS.

- PROVIDE SPACE FOR:

IN-OUT TRAY	STAPLER
DESK TOP FILE	SCOTCH TAPE
ROL-O-DEX	ASH TRAY
CALENDAR	DICTIONARY
RUBBER STAMP	TELEPHONE DIRECTORY
WASTE BASKET	

- PROVIDE STORAGE FOR:

STATIONARY	FORMS
TYPING MATERIALS	OFFICE SUPPLIES
COFFEE CUP	SHOES
PURSE	KLEENEX
LUNCH	PARCELS
COATS (CENTRALIZED)	

- PERSONALIZING WORK STATION:
  PHOTOS
  CARDS   **PLANTS**
  PICTURES   **AQUARIA**
  LAMPS   **NAME PLATE**

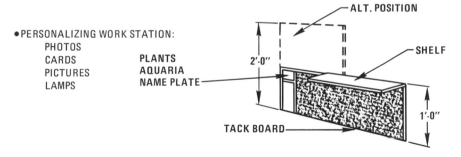

- PROVIDE TWO ADJUSTABLE, MOVABLE, DISPLAY PANELS HIGH ENOUGH TO SERVE AS PRIVACY SCREENS (APPROXIMATELY 4'-6" FROM FLOOR).

- PROVIDE INDIVIDUALLY CONTROLLED LIGHT SOURCE AT EACH WORKING STATION.

- RECOMMENDED PATTERN:

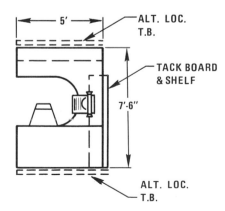

Completed pattern display:

Completed pattern display:

EXECUTIVE WORK STATION

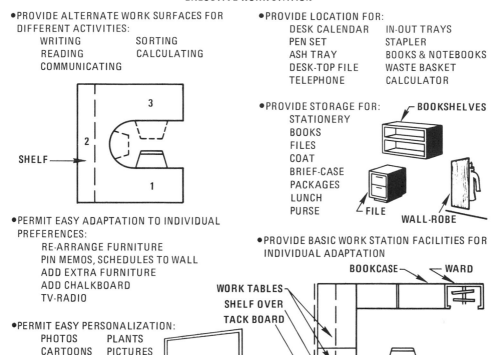

- PROVIDE ALTERNATE WORK SURFACES FOR DIFFERENT ACTIVITIES:
  WRITING   SORTING
  READING   CALCULATING
  COMMUNICATING

SHELF

- PROVIDE LOCATION FOR:

DESK CALENDAR	IN-OUT TRAYS
PEN SET	STAPLER
ASH TRAY	BOOKS & NOTEBOOKS
DESK-TOP FILE	WASTE BASKET
TELEPHONE	CALCULATOR

- PROVIDE STORAGE FOR: BOOKSHELVES
  STATIONERY
  BOOKS
  FILES
  COAT
  BRIEF-CASE
  PACKAGES
  LUNCH
  PURSE   FILE   WALL-ROBE

- PERMIT EASY ADAPTATION TO INDIVIDUAL PREFERENCES:
  RE-ARRANGE FURNITURE
  PIN MEMOS, SCHEDULES TO WALL
  ADD EXTRA FURNITURE
  ADD CHALKBOARD
  TV-RADIO

- PROVIDE BASIC WORK STATION FACILITIES FOR INDIVIDUAL ADAPTATION

BOOKCASE   WARD

WORK TABLES
SHELF OVER
TACK BOARD

DESK TOP

- PERMIT EASY PERSONALIZATION:
  PHOTOS   PLANTS
  CARTOONS   PICTURES
  SCULPTURE AQUARIA

TACKBOARD
SHELF

- VARY OFFICE SIZE IN ACCORDANCE WITH NUMBER OF PEOPLE WHO WILL MEET THERE.

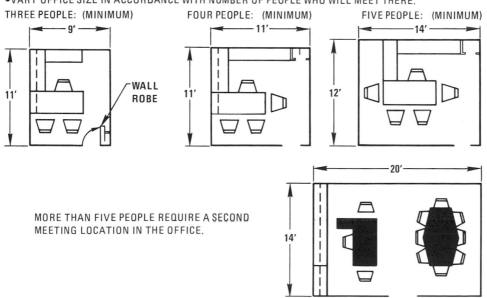

THREE PEOPLE: (MINIMUM)

9'
11'
WALL ROBE

FOUR PEOPLE: (MINIMUM)

11'
11'

FIVE PEOPLE: (MINIMUM)

14'
12'

MORE THAN FIVE PEOPLE REQUIRE A SECOND MEETING LOCATION IN THE OFFICE.

20'
14'

Completed pattern display:

CORRIDORS

- TO PROVIDE ACCESS TO ALL COMMUNITY FACILITIES

  TOILET   WAITING AREA

  STAIRS   DRINKING FOUNTAIN

  LOUNGE   ELEVATOR

- WIDTH TO ACCOMMODATE TWO PEOPLE ABREAST,
  MOVING IN OPPOSITE DIRECTIONS, TO PASS - 7'
  (IN AREAS OF HEAVY
  PERSONNEL CONCENTRA-
  TION, THERE CAN BE
  INTERMITTENT OBSTRUCTIONS)

- MAXIMUM LENGTH WITHOUT INTERRUPTION - 35'

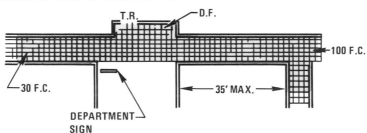

- CEILING HEIGHT TO BE LOWER THAN OFFICE SPACE - 8'

- LIGHTING TO BE NON-UNIFORM, MODERATE LEVEL - 30-35 F.C.
  ACCENT CORNERS AND INTERSECTIONS - 100 F.C.

- FLOOR SURFACE TO BE RESILIENT
  TACTILE CHANGE FROM PUBLIC AREA TO PRIVATE AREA AUDIBLE NOTICE THAT SOMEONE IS
  APPROACHING

- WIDEN CORRIDORS AT MAJOR DEPARTMENT ENTRANCE.

- CORRIDOR SIGNS DESIGNATE DEPARTMENT AREA.

- ACTIVITIES SHOULD BE VISIBLE FROM CORRIDOR.

Completed pattern display:

EXECUTIVE RELATIONSHIPS

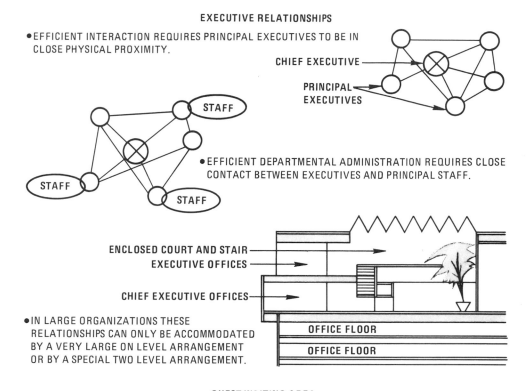

- EFFICIENT INTERACTION REQUIRES PRINCIPAL EXECUTIVES TO BE IN CLOSE PHYSICAL PROXIMITY.

CHIEF EXECUTIVE

PRINCIPAL EXECUTIVES

STAFF

STAFF

STAFF

- EFFICIENT DEPARTMENTAL ADMINISTRATION REQUIRES CLOSE CONTACT BETWEEN EXECUTIVES AND PRINCIPAL STAFF.

ENCLOSED COURT AND STAIR

EXECUTIVE OFFICES

CHIEF EXECUTIVE OFFICES

OFFICE FLOOR

OFFICE FLOOR

- IN LARGE ORGANIZATIONS THESE RELATIONSHIPS CAN ONLY BE ACCOMMODATED BY A VERY LARGE ON LEVEL ARRANGEMENT OR BY A SPECIAL TWO LEVEL ARRANGEMENT.

GUEST WAITING AREA

- SHOULD BE ADJACENT TO ELEVATOR LOBBY AND OPEN TO MAIN CIRCULATION CORRIDOR.

- SHOULD SEAT 10 TO 12 VISITORS.   •SHOULD HAVE A RECEPTIONIST ON A RAISED PLATFORM.

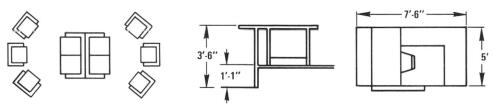

3'-6"   1'-1"   7'-6"   5'

- LIGHT LEVEL TO BE MODERATE, 30-35 F.C., EXCEPT AT RECEPTIONIST DESK.

- AREA SHOULD INCLUDE (OR ADJOIN)
  DRINKING FOUNTAIN   VENDING AREA
  PUBLIC TELEPHONE   CLOCK

- PROVIDE INFORMATION AND DIVERSION.
  DIRECTORY   INTERVAL PUBLICATIONS
  ART DISPLAYS   PLANTS

- RECOMMENDED PATTERN

RECEPTIONIST

ELEVATOR LOBBY

Completed pattern displays:

VERTICAL CIRCULATION

- ●STAIRWAYS AND ELEVATORS SHOULD IMPROVE COMMUNICATION AS WELL AS INTERNAL TRAVEL.

- ●STAIRWAYS SHOULD BE OPEN TO ELEVATOR LOBBIES TO AFFORD AN OPTION
  FOR VERTICAL MOVEMENT.

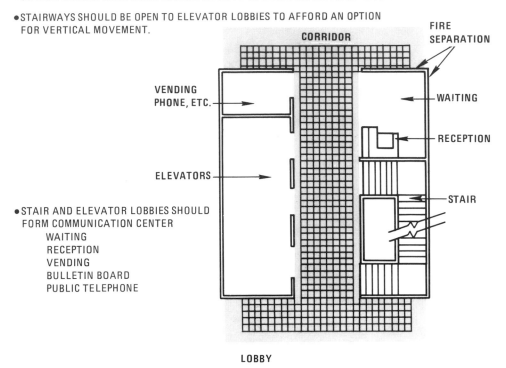

- ●STAIR AND ELEVATOR LOBBIES SHOULD
  FORM COMMUNICATION CENTER
      WAITING
      RECEPTION
      VENDING
      BULLETIN BOARD
      PUBLIC TELEPHONE

- ●LOBBY INCLUDES:
      ELEVATORS
      STAIRS
      DIRECTORY
      INFORMATION (LOBBY GUARD)
      WAITING AREA

  INTERNAL COMMUNICATIONS
      GENERAL BULLETIN BOARD
      PERSONAL COMMUNICATIONS
          FOR SALE
          RIDE EXCHANGE
          ETC.

- ●IMMEDIATE ACCESS TO
      ACADEMIC SENATE
      TRUSTEES AUDITORS
      TRUSTEES MEETING ROOMS

- ●DISPLAY OF THE SYSTEM

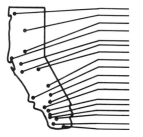

- ●IMMEDIATE ACCESS TO OUTSIDE AREAS OF USE.

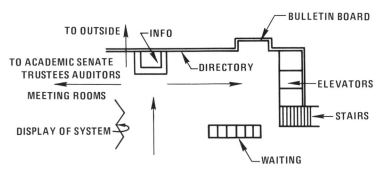

Completed pattern displays:

With a rigorous system the designer is forced to account for all the factors. Without a system he can conveniently forget or suppress some of the less obvious or less pressing ones.

Nevertheless, time is a serious concern for the designer. His mind can work faster than his pencil. As a consequence it is useful to know that while notations in the pattern language form are invaluable records, they are not absolutely essential in the design of a given project. Roslyn Lindheim, a member of the architectural faculty at the University of California at Berkeley, has shown that set theory can be applied to planning problems witout any elaborate organization or the use of pattern statements. While her study of hospital radiology suites was concerned with operating functions rather than social-psychological factors, it illustrates a principle that has wide application.

Professor Lindheim defined her sets as the equipment and the participants needed to carry out a specific step or function in the radiology department. These sets were derived by an uncoupling process. Each aspect of departmental service was separated or uncoupled from its associated activities to determine whether it was an indispensible component of its set and to establish its relationship to any other set. As a result of this process, some activities that have been traditionally seen as inseparable emerge as elements of independent sets. The radiologist, the x–ray image, and the patients' medical records form such an activity set. This activity set has no proximity relationship with patients, x–ray equipment, technicians, film developing, or dressing rooms. The only mandatory tie it has with the rest of the department is a channel or courier for transmitting images and records. With the versatile communication systems that are available, the radiologist is no longer bound to the physical locality of the department. He has been uncoupled from the traditional system.

The advantages that accrue to the designer from this process are substantial. By uncoupling sets, he can define the real relationships rather than the conventional ones. It is a highly productive technique for breaking through the barrier of stereotypes that limits so much of architectural practice, and it can be applied to almost any planning problem. One of the classic sets in formal education the world over is pupil-teacher-classroom. It is the embodiment of the learning process. Yet, when it is subjected to an uncoupling process it rapidly disintegrates. There is no question that the pupil is indispensable. No one has perfected any mechanism by which we can have someone learn for us; it is a completely personal experience. The teacher and the classroom are not indispensable. A pupil can learn from books, films, TV, his peers, and by directly observing and experiencing the world. Furthermore, he can learn in any place or setting where the information source happens to be. As a consequence of this uncoupling, the set can be re-defined as a pupil and a source of information. This new definition of the formal education set was not invented by the uncoupling process, nor is it a statement of anyone's personal philosophy. In order for it to be revealed by the uncoupling process it had to exist. It not only exists, it is widely employed by skilful teachers who encourage independent study and self-generated assignments and who recognize the tremendous educational impact of experience. Yet being able to make this statement in its new form, to define the learning set as a pupil and a source of information rather than a pupil, a classroom, and a teacher, has tremendous implications for both the planning of schools and school systems. Whereas one implies a setting where the student spends his time in the formal confines of a classroom, the other suggests that the only time he needs to be in the classroom is when he seeks counseling or needs direction to new sources of information. A school that reflected this new mode of operation would not only be quite different in a physical sense, it would serve an entirely different function in the community. The results of the uncoupling can sometimes be quite startling.

Yet another interpretive tool has been proposed by Lars Lerup, a Swedish architect and planner who is also on the architectural faculty at the University of California at Berkeley. (Since Christopher Alexander is also on this faculty, that makes a clean sweep for Berkeley in this discussion of translation theory.) Lerup's study was concerned with the development of a process for defining the planning and design goals for residential neighborhoods. After carefully analyzing the characteristics of three typical housing solutions—the single-family home, the row-house, and the apartment—he prepared a series of Pattern Statements about each of these characteristics using the notation developed by Christopher Alexander. In order to evaluate these characteristics and select from them the set of characteristics that would most closely meet the needs of the anticipated users, he employed a form of matrix called a morphological box or a "Zwicky Box," after the Swiss astronomer, Fritz Zwicky.

Using the box is extremely simple, though developing the information and the judgments that are displayed in the box may be very time-consuming. Assuming that a set of ten criteria have been chosen and analyzed, these can be listed along the left margin of a sheet of paper. Across the top the three housing types to be compared are listed. By projecting lines down from the three housing types and across from the criteria a matrix is created. Each line intersection represents a characteristic of one of the housing types. With the analysis of each of these characteristics in hand, a prospective user can start making selections by checking the selected intersections.

He may, for example, prefer the low maintenance characteristics of the apartment to the high maintenance characteristic of the single-family home. He may prefer the image of the single-family neighborhood to any other. He may opt for the higher degree of visual privacy afforded by the row-house. As these selections are made by circling the appropriate line intersections a desire line is generated that defines a housing solution that is best adapted to his needs.

It is obvious that everyone will not generate the same line. The accompanying diagram projects the desire lines of three different family types on Lars Lerup's matrix: childless young couples, couples with children, and childless elderly couples. These are hypothetical lines, developed solely for illustration purposes. Actual lines could only be plotted for these or other groups through their direct participation.

As you can see, the process results in an uncoupling action. By selecting characteristics from a variety of housing types we have removed them from their customary settings and have defined a new housing type. At first glance this seems like the process used to create the assembled animals of mythology, those with the legs of a gazelle, the neck of a camel, fangs like a lion and the hide of a leopard but that is not a fair analogy. Nature's designs are highly evolved to fill an ecological niche. The standard patterns of housing development are not. Indeed, from a social-psychological standpoint, if the desire line deviates at all from the prototype it is indicative of some fundamental mismatch in the prototype.

It is obvious that in responding to a desire line—creating the camelopard—the designer cannot allow himself to be constrained by stereotyped images. If the line shows a preference for the high identity of a single-family home and the lower land taxes of a row-house, the designer cannot meet those criteria simultaneously and still maintain the traditional front, rear, and side yards that are typical of suburbia. If another line shows this same preference for the single-family identity but links it with low-maintenance apartment living then the designer is obviously faced with the task of improving the identity of the individual apartment dwelling. Or, to be more precise, he must abandon these stereotyped names entirely and create new forms that reflect living preferences.

The fundamental virtue of any systematic method for analyzing and evaluating design data is its creative potential. Whether it is Pattern Language, uncoupling, the morphological box, or any other process, the organization of information they provide offers the designer a means of identifying real problems and, consequently, devising better solutions. These methods are a great help in breaking through the stereotype blockade that limits the range of traditional solutions, and can open the way to true innovation—not the pointless innovations of form and style—but those that are relevant to human needs and concerns.

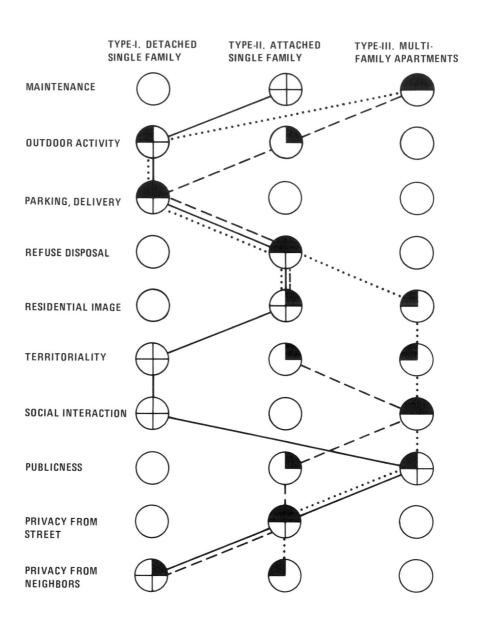

It is true, of course, that analytic and evaluative systems offer nothing more than an aid, an opportunity. They do not automatically provide solutions. The designer must still perform the crucial operations of synthesizing the information and defining the solution. His ability to realize the potential benefits of these procedures will ultimately be limited by his ability to free himself of the preconceptions acquired during a lifetime of training and practice. He can't make the automatic assumption that all of the conventional forms and practices of architecture and planning are unsuitable. Some are and some are not. He must maintain an open attitude toward all solutions and be as free to accept a conventional answer when it satisfies the need as he is to abandon it when it no longer works.

Side-Stepping Stereotypes

Throughout this discussion the word "stereotype" has been repeatedly used as a term of opprobrium, a practice which is not wholly fair or accurate. Since the problem of stereotypes or pre-conceptions is such a crucial matter in the design field, as it is an any area of human endeavor, the topic deserves more detailed consideration.

As it is used here "stereotype" can be defined as a standardized mental picture held in common by members of a group that represents an oversimplified opinion, or an uncritical judgment of some race, institution, issue or event. The number of these pictures that exist in any human mind must be enormous. They range from ideas about the attributes of the founding fathers of the republic to the characteristics of other races or social groups. They are treacherous material to rely on in arriving at judgments but they persist because they offer a great economy of mental effort. Furthermore, it is not always possible to distinguish between a factual picture of the world and a stereotyped picture of the world on the basis of the information at hand. A judgment or decision based on a careful evaluation of available information is not a stereotype. If, however, the same picture persists after later information indicates that it is no longer appropriate, it becomes a stereotype. Fossil ideas like this that have been undermined or set adrift by new data float like stately icebergs through every field of human activity, magnificent in appearance but wholly without foundation.

The great attraction of stereotypes or conventions is that they relieve us of the necessity of making difficult decisions. For the designer, the mind-wracking problem of analyzing human behavior and originating specific solutions is largely by-passed if he adopts a conventional solution. It is also a safe procedure. A design that falls within the current mode or fashion is buttressed by precedent, a comforting crutch in the face of uncertainty.

Thinking in patterns of convention is a great time-saver. It elimi-
nates the necessity of reinventing the wheel every day. There are a
host of considerations that every designer deals with in conventional
ways in order to save his energies for those aspects of the problem
that he considers more important. If he stopped his work to re-
evaluate the design of every doorway to determine if there is some
better way to pass from one space to another, his work would never
end. In accepting these convenient short-cuts, of course, he runs the
risk of employing stereotypes where they are ill-suited and concentrating his attention on problems that don't need to be solved.

One of the curious virtues of stereotypes is that they form a substantial part of the raw material of thought. The idea that we should
free our mind of preconceptions in approaching any new problem is
laudable and understandable but as a practical matter it may be
impossible. They form a part of our supply of action-schemes, the
responses we have learned to employ in dealing with the complex
problems of living, and as Jean Piaget has demonstrated, these can be
modified by a process of accommodation but not eliminated. In this
sense, it is more accurate to say that we should open our minds to the
process of accommodation rather than to eliminate our preconceptions. One statement reflects the actual nature of thought modification as the other does not. It also reflects some of the benefits of social
evolution rather than social revolution. As Christopher Alexander
has pointed out, tradition or continuity makes a system or a culture
viscous; it tends to dampen out wild swings from one extreme to
another. The practical implications of his observation are obvious. If
we were to awaken to an entirely new world each morning none of
us would know what to do with it.

One demonstation of the powerful hold conventions have on our
minds is our tendency to use them as analogues in what is intended
to be original thought. This is a particularly popular exercise in the
design field. A designer may characterize his irregular building form
as being "like a village," or he may describe his school plan as a
"shopping center for ideas." In each case, he is employing the title
of one stereotype in an effort to imply new qualities in another
stereotype without much concern for accuracy. Having done the
same thing myself on countless occasions I am aware of both the
benefits and the shortcomings of this procedure. In evolving new
configurations in response to human behavior needs, the problem of
finding adequate labels becomes severe. Calling new devices or furnishings "contraptions" doesn't communicate any information at all.
Calling them "upholstered bleachers," or "lounge queues" may be
applying stereotyped labels in a wildly inaccurate way, but at least
the terms convey some image of a familiar form or activity.

This name problem became particularly difficult in our Cal State study. The terms "student center" or "campus union" refer to a building type that is well-known in academic circles. Even people who have never seen such a building probably have some vague image generated by the name itself. Since we hoped to sidestep the stereotype problem and deal directly with behavioral and social-psychological factors, unadulterated by preconceptions associated with this building type, we attempted to avoid the use of these terms completely in pursuing our study. Unfortunately, it didn't work. Attempting to define the focus of our study without using the common nomenclature became impossibly wordy. It was confusing to everybody involved, including us. Yet, when we retreated to the use of the term "campus union," communication was instantaneous. As I said before stereotypes are a great time-saver.

In view of the persistence of stereotypes and their inconspicuous way of imbedding themselves in our minds, it is necessary to redefine their role. They are best described as unreliable crutches that must be constantly tested and inspected. We cannot abandon them as we might like to do since that would mean we would be completely immobilized. It is a strange symbiosis; we can't trust them, but we can't get along without them.

In the field of planning and design, Kevin Lynch is one of the few who have recognized this peculiar relationship. "We cannot do without these stereotypes. The process of city design would become impossibly tedious for lack of such crutches. It is the use of such models without reference to purpose that is defective as well as the poverty of our stock."

All the foregoing may seem like a general *apologia* for stereotypes and a defense of the narrow mind, but that is not the point. It seems sensible to me that before embarking on a crusade against the Philistines we need to understand their nature. As the architect faces the problem of translating abstract design criteria into three dimensions, it is useful to know that it is a natural tendency to rely on his personal stock of conventions in searching for a solution. This knowledge at least permits him to examine such conventions before use to make sure they fit and it increases the possibility that he will recognize the need for new, non-conforming, answers. Being conscious of the process that leads us to keep repeating ourselves is our best safeguard against the tyranny of habit.

It is true that the patterns of thought and the inventory of forms that are the result of learning and experience impose a formidable barrier to original thinking. Even when the designer is conscious of these limitations there is no easy way to insure that he will surmount them. This seems to be particularly true in dealing with social-psychological data. Translating such information is often extremely puzzling just because none of the conventional answers exactly fits these abstract criteria. They sometimes, in fact, lead to bizarre conclusions that have nothing in common with anything the designer has ever seen or envisioned. It is always possible, of course, that at such a point the designer is simply pursuing some mad delusion, but it is equally possible that he is making a creative breakthrough. Such a result can easily precipitate a crisis of self-confidence. Yet, if the designer has used some systematic process for marshaling his information and has consciously evaluated the design consequencies of each factor he should be able to accept the result not just with equanimity but with enthusiasm. The crisis, if it occurs, stems from the difficulty of accepting the facts of human values, goals, and strains as the true criteria of design. The old familiar forms are so much more comfortable and reassuring.

* * * * * * * * *

In the long chain of events that leads from the inception of a planning or design project to a completed structure or a realized community the translation phase is only one episode. It is the crucial episode, however, that determines how well the solution fits the problem. In view of its importance, it is puzzling that in the long history of the design professions, the translation process is only now beginning to receive the searching analysis it deserves.

Translation always occurs, of course, with or without a system. Even when the designer selects some convenient stereotype, a "house," a "church," or a "suburb," and stuffs a set of human activities into its confines it is a translation even though it may be woefully inadequate to accommodate these activities. Even when the solution is adequate, or better yet, outstanding, it is generally considered to be a result of intuition, creative talent, or even genius. In view of the focal role of translation in the creative process this approach is entirely too haphazard. Some systematic process such as those that have been discussed in the proceding pages is essential.

These systems of translation are not in themselves unique inventions. It is probable that the creative mind has always used some form of uncoupling to generate new solutions. The creative mind is perfectly capable of employing a morphological box or using pattern language without committing them to paper or even knowing that these devices have a name. What these systems do is to extend greatly the range of our mental capabilities. While we can easily juggle half a dozen criteria without any props, real problems commonly have scores or even hundreds of criteria. Without some method of organizing this information so that it can be consciously considered, the designer either will be immobilized by an information overload or will subconsciously suppress all but a manageable number of criteria. In either of these cases he is not solving the whole problem.

One realization that highlights the critical importance of the translation phase is that once the process is completed the solution has been defined. It may not look like anything that anyone has ever seen before but it is, nevertheless, the answer. The steps that still remain to bring it to realization may be extremely complicated and time-consuming, but they are subordinate and supportive. The designer must select or evolve a structural system, select materials, and develop utility systems within the constraints of budgets, codes, and the realities of the construction industry. He must, also, make a host of decisions about design characteristics that may not be implied by his criteria; form, color, texture, and proportion. In dealing with these subordinate issues he has a wide range of options and great scope of ingenuity. The thing he must constantly guard against is the temptation to let the final result drift away from the original concept due to a preoccupation with some novel structure or some preconceived form.

When an architect or planner has gone through the procedure of gathering the necessary social-psychological data, defining the requirements in human terms, and translating this information into three dimensional statements the solution has been defined.

Chapter VI

A New Definition
Of Design Quality

The practicing architect or planner will probably have some vociferous objections to my flat assertion that once the translation process is completed the problem has been solved. In the normal course of practice this step would be considered only a bridge between a written program and the beginning of design studies. It is these later stages (which I described as subordinate and supportive) that have long been considered the focus of the creative process. Problem solving, per se, is not a highly regarded ability in the design field. It is the art of reforming the problem into an esthetic statement that captures the attention and the applause of architects and designers.

It must be obvious by this time that I hold a different view of the priorities that are appropriate for the design professions. There is a vast difference between solving the problems of human use and letting these solutions evolve an esthetic of their own and the alternate process of forcing these same uses into a fashionable form. Reforming the problem into an esthetic statement is a process of taking criteria of human use and manipulating them in accordance with a second set of esthetic criteria that may have nothing to do with the users. This can easily lead to solving only selected aspects of the problem or ignoring it altogether. Constance Perin has commented on this tendency with the allegory of the bus driver who comfortably solved his own schedule problem by the simple device of not stopping to pick up passengers. This kind of problem solving is distinctly not needed in planning the human environment.

It is not my intention to venture into the embattled field of architectural esthetics. It is already littered with the bodies of countless writers who have attempted to define "good" design or to describe a process that would ensure design quality and have all, in time, been cut down by a withering barrage of critical fire. It is a field that is distinguished by a particularly convoluted prose that the English writer Rupert Spade has characterized as "Semantic Drunkenness," an elaborately structured defense mechanism against the dreaded cry "the Emperor has no clothes." My limited objective, in line with the general concern of this book, is to discuss design quality as it relates to human behavior.

As some of the quotations in the first chapter indicated, a number of behavioral scientists have been sharply critical of what they consider to be an undue architectural emphasis on appearance at the expense of the more fundamental problems of human use. The essence of their argument is that, over the long term, the elements of the environment that have positive value are those that relate to fundamental human striving rather than to outward form or surface appearance. In their view the design professions have been preoccupied with their own artificial values and goals and, in seeking to create unique and memorable architectural forms, have solved their problem like the bus driver, by ignoring the legitimate concerns of the users. In support of this indictment they offer some meticulously conducted studies that demonstrate how ludicrous some architectural environments are when evaluated in terms of human use. Reading the work of such careful investigators as Robert Sommer and Sim Van der Ryn is an endless embarrassment to a concerned architect. It is hard to understand how the designers of the settings they studied could have missed the point so completely. These studies induce a feeling similar to the bewilderment expressed by Lord Macaulay when he commented on the preposterous conclusions arrived at by the brilliant minds of medieval philosophy: "They showed so much acuteness and force of mind arguing on their wretched data that one is perpetually at a loss to comprehend how such minds came by such data."

Accepting at face value the criticism of the behavioral scientists concerning priorities in design would lead to the conclusion that the appearance of the world we live in is a matter of no importance, a view that the architect would have the greates difficulty in accepting. Even if he were willing to ignore the judgments of his peers, he is faced with incontrovertible evidence that appearance design makes a difference. In the expansive spread of human choice ranging from shopping for food to the selection of a home, people consistently base their decisions on form as well as content. In many instances it is form alone, or the complex or color, form, and texture that defines surface characteristics, that dictates the choice. It is the flawless red apple that commands attention in the market place; the question of its nutritional value and flavor may never be raised.

Even more compelling evidence, from the architect's viewpoint, is supplied by the reaction of the one class of people who are indispensable to his existence—his clients. There is no question in my mind, and I believe most architects would agree, that many clients, whatever their ostensible show of objectivity, in the absence of personal factors will base their selection of an architect on the appearance of his work. While it is true that this is a quick and easy basis for judgment whereas a thorough analysis of the actual work would be very tedious and time-consuming, the importance of our emotional response to physical forms cannot be discounted. There is no consistency guiding this choice. One client may favor striking or novel design features that another would find abhorrent. Yet both are responding to external appearance.

The insistence of behavioral scientists that the essence of design for human use is found in the content of the solution and the clearly demonstrated tendency of the human users themselves to be influenced by the form of the solution presents an apparent dilemma. The two are not necessarily incompatible, of course, and in theory, a perfect solution would serve both ends perfectly, putting "context and form into frictionless coexistence" in Christopher Alexander's words. In my view this is not an impossible goal so long as it is qualified in one crucial sense; content in terms of human use can be in perfect harmony with the design qualities that influence our responses *as long as both are judged from the user's point of view.* It is when we adapt or restate, or reform the content to fit the special values of the design professions that the glaring mismatches on which behavioral scientists love to dwell occur.

At first glance, the idea that the user's point of view is the only valid basis for evaluating design quality would seem to relegate the determination of esthetic values to the public opinion polls, but the distinction I am trying to draw is much more subtle than that. It is my contention that design quality, which we can consider to be the way a thing looks in contrast to the way it works, is a motivating force in our society, but that it should be judged as a functional factor rather than a matter of the designer's personal taste. In other words, as the designer proceeds from the basic solution provided by the translation process, the decisions he makes about form, materials, color, texture and detail should be made in terms of *what these choices communicate to the user* rather than his own philosophy or the fashion of the moment. It can be argued that the gifted designer does exactly that and, in addition, opens new experiences and new visions for the user. That may well be his intention, but in considering architecture as a communication medium, it sometimes appears that the course of design seems to be directed deliberately toward mystifying rather than communicating.

How We Move Determines What We See

One aspect of perceptual psychology that has important implications for design is the study of the differing perceptual consequences of exploratory and habitual movement. The idea that architecture can only be experienced fully by movement is well known, of course, and stems from the fact that any large form can be accurately perceived and understood only as the viewer moves through and around it. This is the concept adopted by some critics as a criteria for evaluating architecture and has lead to such sophistries as the idea that public places should be choreographed as ballet movements, two steps up, a glissade to the left, and two steps down.

While the realization that we perceive or "read" architectural settings through movement is an important and useful one, it is unfortunate that the attention of architectural theorists has not gone on to a fuller understanding of this phenomenon. Taken by itself, this evidence suggests that any viewer or participant in an architectural setting is constantly reacting to a series of changing stimuli in response to the forms around him. Nothing could be further from the truth. The fact is that, for the most part, the environment we are in is largely an indistinct background.

This apparent contradiction stems from the fact that we habitually employ two distinct forms of locomotion, each with distinct characteristics and each with different consequences so far as perception is concerned. As Robert Bechtel has described this phenomenon, whenever we encounter a new environment we move in an exploratory mode, sensing, scanning, and probing. In a sense, our actions are similar to those of other animals who must investigate their surroundings before settling down to relax and could stem from a far gone day when human beings inhabited natural environments in which it could be fatal to enter unknown territory without due caution.

Once new territory has been explored, however, we shift from an exploratory to a habitual mode of movement and our personal concerns become paramount. Our surroundings fade into the background and we move with a purpose, intent on our destinations or our private thoughts. Substantial changes can be made in a familiar environment without being noticed unless they affect the landmarks we use for navigating to our personal goals. That may seem like a dubious statement but it is convincingly supported by a study at the

University of Michigan on the effects of windowless classrooms. The study was intended to measure changes in student attitude or performance resulting from the elimination of windows in the classroom. The results indicate that the shift from one condition to the other was not always noticed (which raises the question—how much change would be required to be noticeable, taking off the roof?).

During exploratory movement we are most open to whatever message architecture has to communicate. We are not only receptive but actively looking for clues that will tell us where we are, what the surroundings offer us, and what behavior is appropriate for the moment. Once our curiosity has been satisfied, and it may take only a brief time, it will be very hard to arouse again. The details of the environment have become embedded in an environmental *Gestalt*.

In view of these facts, it appears that the architectural profession has not overestimated the *importance* of design quality; it simply has not considered it with sufficient precision. Design characteristics are important to different people at different times for different reasons. What is most important is that what is communicated by the design is what each viewer needs or wants to see at the moment, not what the designer hopes to have him see.

The reason for some confusion on the importance of salient design features is not hard to understand. Most of the evaluation of architecture and urban design is made by writers and critics who fall in the category of visitors to the scene. As a consequence, the literature of architecture is dominated by the responses of those who are seeing and experiencing new environments in an exploratory mode. As a group, they are highly sensitized to characteristics that are largely invisible to the customary users and to any non-professionals who may be experiencing the same setting for the first time. As a further distortion, these reporters labor under the burden of having to report something, (an obligation that must sometimes be difficult to discharge). Designers rarely have an opportunity to hear a report from the people who use a building or a townscape as a necessary adjunct to the process of living or even from a lay visitor. As a consequence the critical literature of architecture exhibits gross distortions; it is not only focused on the abstractions of design quality at the expense of living quality, but it deals with these issues as though they were the exclusive province of the designer. If design is to communicate anything, then it must be couched in a language that is comprehensible to both sender and receiver. Any communication that persistently ignores the receiver obviously is intended to maintain a communication block.

What Does Architecture Communicate?

There are a host of design elements in the American city that serve as communication symbols. The most abundant, of course, are the ubiquitous signs that direct our movements and vie for our attention, but they are by no means the only ones. Though it is not customary to think of them in such terms, building entrances, stairways, and street corner curbs communicate also. Even though they may vary greatly in detail, each of these concepts is so embedded in our routines that their message is seldom misunderstood. If we are looking for a way to move between levels, the sight of a stairway communicates all we need to know. There are seldom any signs explaining why we should wait at the curb before crossing the street but we are all perfectly aware that standing in the gutter, only inches away, is infinitely more hazardous than staying on the curb. The multitude of conventions that fall into this category are so much a part of our way of life that we seldom consider them as communication symbols or design elements. They are part of the web of our culture.

There is another category of design elements that function as communication symbols in a way that is much harder to describe. This category includes the characteristics—discussed in an earlier chapter —that provide us with clues about the suitability of a given environment or establishment for satisfying our immediate personal needs. We make all manner of choices on the basis of such clues. From a design point of view these clues must be considered the most positive motivational effects of architectural form yet they have received almost no systematic study. The subtle distinctions that cause any one of us to turn into one motel rather than another, or to select one apartment building in preference to another, have not received the attention they so clearly deserve. Anyone who hopes to employ this category of clues as symbols for communicating intentions architecturally is faced with the substantial problem of evolving his own understanding of both vocabulary and grammar.

The architect's problem is much like the problem faced by a linguist who is attempting to understand a language that has never before been analyzed. Design elements exist as a communication medium just as the language exists as a communication medium. The problem the linguist faces is to make sure that he identifies the nature of this special language rather than interpret it in terms of other languages he already knows, to be certain that he understands what it means to the people who use it rather than what it means to him. The architect is faced with the same kind of problem: finding out what the forms, colors, textures and materials he uses convey to the people who are expected to respond to these characteristics. It is very easy to go wrong on this point. When a structure is successful in communicating, it is easy for a designer to assume that it is in response to the interest or attraction generated by his design. In fact, the communication may be at an entirely different level; the man on the street may simply be reacting because the structure generates an image of newness or accessibility.

In suggesting that the puzzle of how design characteristics communicate can only be understood by studying the reactions of the receiver rather than concentrating on the intentions of the sender, it may seem like the description of a closed or static system, but this is not the case. The fact that that the communication capability of design is limited by the vocabulary and syntax available to the viewer doesn't mean that this limitation is forever fixed. It can be changed as any language changes, but only in discrete steps and over a period of time. A writer who wished to address a message to the American public would never think of employing a foreign language. If he did, it would be a certain sign that he did *not* want to communicate any direct message to this audience. He could. however, introduce new words or terms from other languages into his text without confusing his readers if he did it in a manner that would permit the reader to infer their meaning from the context. By this process the reader would actually expand his vocabulary.

Exactly the same process applies in the design field and the field of art as a whole. In the history of art and architecture there have been enormous shifts of style and symbolic content without any impairment of communication because the changes occurred in an evolutionary manner over considerable periods of time. The current fashion in these fields is to originate entirely new styles overnight. So far as comprehensibility and communication are concerned this tendency produces the same result as if the designer suddenly elected to speak in Urdu. There is *no way* the viewer can grasp the faintest glimmering of what the message is about. He would have no base point from which to begin. In view of the nature of communication it must be assumed that designers and artists who elect this idiosyncracy do not really want to communicate to the general public. They are aiming their messages at a closed circle of their peers who speak the same secret language.

While there has been no systematic effort to decode the symbolic content of the elements of our environment as they are read by the layman, there are studies that rather clearly demonstrate that layman and architect attach different meanings to these symbols. Using photographic measurements for eye pupil size as an index of interest or arousal, Ifan Payne, of the Polytechnic in London, has studied the response of architects and non-architects to color slides covering a range of architectural and non-architectural subjects. His report in-

dicates that while both groups react, they react to distinctly different
elements, suggesting that the two groups have different responses to
the same stimuli. Robert Hershberger, working at the University of
Pennsylvania, made the same comparison in a more elaborate way
and came up with even more conclusive proof of a wide communication gap. Working with architectural students and non-architectural
students, he asked both groups to respond to colored slides of buildings on a set of semantic-differential scales, a device that permits the
respondent to indicate a wide range of emotional responses to graphic stimuli.

Both groups with which Hershberger worked responded emotionally to building characteristics, which indicates that there is some
basis for assuming that architecture does communicate; there was,
however, a stunning disparity between their interpretations.

> *Approximately 30% of the time when Penn architects would judge a building
> to be good, pleasing, beautiful, interesting, exciting, and unique, the non-ar-
> chitects would judge it to be bad, annoying, ugly, boring, calming, and common.
> Such a large number of differences between the two groups would, of course,
> seriously affect the success of the architect in communicating his intentions to
> laymen.*

I love the calm understatement of such reports. When two groups of
people employ a language in which thirty percent of the critical
terms are not only interpreted differently but actually have opposite
meanings, communication is not just seriously affected—it is completely blocked. What is even worse is the prospect that the message
will not merely be misunderstood, but that it will convey an opposite
meaning. There can be no hope of communication between architect
and layman through the medium of form while such a disparity
exists.

Robert Hershberger concludes his report with an analysis of the
options that are available for improving communication through the
medium of architecture. One option would be to educate the American public in the special meaning architects attach to form, a truly
Herculean task because of the propensity of designers to generate
new forms at a fantastic rate. A second course would be to re-orient
architectural education, so that it doesn't alter the lay perceptions the
entering student brings with him. Since much of architectural education is concerned with altering the students perceptions and establishing a new value system, this proposal would, in effect, call for the
establishment of a non-architectural education system. The third
possibility is to re-orient the training of a young architect to create
an awareness of how forms are interpreted by laymen. This last
prospect is the one that seems both reasonable and productive. It
would make the architect bi-lingual, with one vocabulary for address-
ing the public and another for communicating with his peers.

Before leaving the subject of architecture as communication, it seems worthwhile to attempt some explanation of how the gross distortion between the message the designer intends to convey and the message that the viewer receives comes about. It is the result of a lack of feedback. When we communicate in verbal terms and our statement elicits a proper response or an appropriate action, our assumption that we employed the right terms is reinforced. If our statement elicits an improper response or an inappropriate action, we have to re-structure our statement until it produces the desired result. It is this process of feedback, the reaction of others to the things we say, that keeps our communication skills sharpened and in tune with the rest of the world.

Unhappily, there is no similar process at work in the design field, or, to be more accurate, the design field has never established and maintained channels that would accommodate a ready flow of feedback from the lay public. The feedback the designer gets is largely from his peers or from the design press, a highly specialized reference group which tends to reinforce conformity to group values and to give negative responses to any deviation from them. If the only feedback you get comes from one specialized group, there is a strong tendency to keep altering your statements until the feedback is positive. The net result is that this kind of closed system not only develops a special language, it also exercises strong sanctions on its members. After a time there is no one else you can talk to.

Christopher Alexander has discussed feedback from a somewhat different perspective. It is his view that only in primitive societies— where architect, builder, and user are one and the same—that anything approaching perfect feedback exists. In primitive circumstances, as users alter and modify their environments over time, typical building forms emerge that are specific to a given area and a given way of life. Since they are the result of continual feedback from the users, who have only to act to correct a deficiency, the resultant form is beautifully adapted to both the available resources and the existent human needs. When an individual in this kind of society needs a house, he doesn't need to ponder the question of what kind of house; the term "house" is totally descriptive of material, form, structure and process.

Only slightly different from the primitive process are the relationships and feedback channels in what is called "vernacular architecture". In this case the user works with a designer-builder in the construction of a building type that is common to their area and well known to both of them. The barn builders who erected such handsome and durable structures in the eastern United States during the last century were working in a vernacular style. While the details might be varied to suit the preferences of an individual user, or to express an individual builders originality, the nature and purpose of these structures were widely known and they were generally similar in design.

Alexander describes the architecture produced by primitive societies as "unselfconscious" in contrast to the "selfconscious" architecture produced by societies that regard architecture as an art form that is the special province of architects. Unselfconscious forms evolve in response to direct feedback from the users and become an integral part of the culture. Selfconscious form-making, on the other hand, is something that is done for the user, with little if any participation on his part. Furthermore, the user often has no opportunity to register his views about the result through direct feedback. The apartment house tenant can complain to the building superintendent, who may or may not relay this to the building owner, who may or may not tell the designer. With such a complicated and uncertain feedback channel, modification of the system to provide a better reflection of user needs takes a long time if it occurs at all.

There is no point in wistful longing for a simpler day when each of us might have controlled his environment through direct means. Our tendency to herd together in great urban centers demands an increasingly sophisticated technology to solve our common problems. This in turn leads to specialization both in the design of buildings and in the planning of communities to deal with needs that are totally beyond the scope of unselfconscious societies. That doesn't mean that we have to settle for impersonal environments that are beyond our control; the basic change that is required in our present methods is the introduction of a direct feedback loop in the design process that would make designers more responsive to the human needs of users. Given the scale of our problems and the number of people who are involved, the processes of the human sciences that have been described in the preceding chapters offer the best hope of doing that.

Feedback from users is the missing element in selfconscious architecture and planning. It is not only indispensable in correcting the content of solutions and producing a better fit between the problem and the answer; it is the only possible way to arrive at a language of architectural form. Regardless of the designer's talents or intentions, unless he has an audience that speaks the same language that he does there is no communication. Communication is a two-way street.

Design Principles (of Human Origin)

In wrestling with the problem of what constitutes architecture or "good design," the Danish architect Steen Eiler Rasmussen came to the conclusion that, whatever it is, it is beyond words.

Architecture is not produced simply by adding plans and sections to elevations. It is something else and something more. It is impossible to explain precisely what it is—its limits are by no means well-defined.

Rasmussen is by no means the only one who has arrived at a similar impasse in searching for the essence of architectural merit, though he is one of the few who have been candid enough to admit it. Other writers are prone to take refuge in a smoke screen of words or a retreat to mysticism. As a result, the literature dealing with architecture as an art tends to deal in mystical pronouncements such as "God is in the Details," or the convoluted prose Rupert Spade has described as "Semantic Drunkenness." In reading this literature, it is hard not to equate it with stories of the medieval alchemist's search for the philosophers' stone, that miraculous instrument that would transmute base metal into gold. The only reason for assuming that the philosophers' stone existed was that the alchemists so desperately wanted it to exist.

In view of the difficulty of arriving at a magical formula for great architecture, it is not surprising that the profession would ultimately settle for more distinct and more measurable criteria. The venerable maxim, "Form follows function" would be one of these. The external expression of the structure of the building, and the manner in which it is adapted to the natural site are other criteria of the same nature. Ideas like these have the attraction of being easy to grasp and easy to apply but they do not stand up very well under close analysis. Just what function is the form supposed to express? In one widely publicized building the function the architect elected to express was the ventilation system. What the resulting series of giant smokestacks was supposed to communicate to the public is not clear, but it could hardly have much significance in terms of human concern. This kind of exhibitionism substantiates Christopher Alexander's observation that, in practice, forms tend to reflect the most easily grasped issues. I might add that they also tend to reflect those factors that produce the most striking and novel forms.

Since design quality is difficult to describe in words or to reduce to formulas, another course that can be followed is to select examples of excellence that can serve as models. This, in fact, is the course that has been most consistently employed by the design professions. In one form or another, they all sponsor some kind of award program as a recognition for outstanding work. The American Institute of Architects has conducted an annual program of awards for outstanding architectural work for many years. In this Honor Awards program, a selected jury considers photographs of buildings submitted by architects and selects those they consider to have particular merit. The resulting winners are widely publicized and serve as models, both to the public and to the professions, of what good design looks like.

There are several drawbacks to this system. For one thing, judgment is made on the basis of pictures, not buildings, a process that has lead to considerable distortion and artifice. One photographer who specializes in photographing buildings has argued that awards should be made jointly to the architect and the photographer, a point of view that is not altogether unreasonable in view of the striking difference that can sometimes be distinguished between the picture on which the award was based and the building itself. When, as has happened, the profession honors a building that the users consider a disaster, it cannot help but raise a serious question of the credibility of architectural judgement. In the aftermath of such episodes, the Honor Awards program has been revised so that each building selected must be personally visited by at least one of the jurors. While such a visit can hardly permit the kind of serious analysis that should be the basis for judging the merits of any setting for human activity, it at least weeds out the obvious failures. A more basic challenge to the awards program is that it simply reflects the passing parade of fashion rather than the fundamental characteristics of outstanding architecture. This argument is not easy to follow without access to the awards of earlier years. It is only when the photographs of the award-winning buildings are viewed in sequence that the remarkable shift in architectural styles that has occurred in the past forty years can be fully grasped. Seen together, they create the distinct impression that architectural styles have changed at such a rapid rate that they are perilously close to changes in the field of fashion or the annual models of automobiles rather than an art form rooted in immutable principles. They create the impression that the paramount concern shared by the form givers of the past few decades is a ceaseless search for novelty and innovation in form.

The organized profession has taken the position that, while good design may not be definable, it is at least something that able architects can recognize. In light of that belief the most humiliating experience the profession ever suffered occured in the design of the American Institute of Architects headquarters in Washington, D.C. In an effort to achieve the kind of outstanding demonstration that would be appropriate for the home of a national organization of designers, the Institute sponsored an open design competition, judged by a carefully selected jury of ranking architects. Not one, but several designs were selected in a preliminary round, and these winners were commissioned to amplify and develop their concepts in order to demonstrate their full potential. The single winner finally selected had thus survived the rather rigorous ordeal that the profession regards as an appropriate method of selecting architects for assignments of great public moment. It was an ideal demonstration of an ideal method.

Unfortunately, the profession had previously planted a quietly ticking time bomb that was ultimately to explode in its face. In view of the importance of the nation's capital as a functional and symbolic center, they had supported a design review commission of prominent laymen and architects with considerable power to approve new developments in the Federal center. The need for an agency to exercise some control over the rather haphazard growth in this focal point of national life is clearly apparent to anyone who knows Washington. As a consequence, the design for the national headquarters for the architectural profession, arrived at through a model process, had to be submitted to the Fine Arts Commission, a model design review agency, for approval. Approval was not forthcoming; the inconceivable occurred and the headquarters design was rejected by the commission. The events that followed, the series of modifications that were proposed and rejected, the resignation of the architectural firm that had won the competition, and the appointment of a new firm that developed an entirely new design that was finally approved, are outside the scope of our interest. From our point of view, the only purpose for reviewing this episode is that it so clearly demonstrates that the elusive element called "design quality" is impossible to define and obviously does not mean the same thing to all architects. Rather than reflecting a consistent set of ageless principles, good design is partly a matter of personal judgment and partly a matter of the constantly shifting patterns of taste in our culture.

In view of the difficulty the profession has experienced in arriving at a stable definition of good design through formulas, criteria, or even by example, it would seem to be time to consider an entirely new approach. In line with the principal argument of this book, it would be appropriate for the profession to shift its emphasis from a consideration of architectural quality in terms of form to the much more difficult question of suitability for human habitation. As Robert Sommer has pointed out, a design problem is a value problem. It is a question of whose interests are being served. From a humanistic point of view, this must mean that the interests of the users, those who experience the building most intimately as residents or workers, those who visit it or enter it as consumers, and those who experience it only vicariously as part of the community scene must be served first. There are other valid points of view, that of the investor, the constructor, and the community officials who are responsible for public health and safety. Nevertheless, any method of evaluation that ignores the users evades the principal issue.

Describing a set of criteria for defining design quality in humanistic terms produces a list of characteristics that is surprisingly short and, at first glance, hopelessly vague. They are so completely at variance with traditional criteria that it may be difficult to perceive that they deal with the same matters. To understand them it is necessary to discard, or at least to hold in abeyance, many of the values that are deeply anchored in professional value systems.

The first of these criteria is that human values are the primary concern of design. This does not mean the kind of value statements that are so often quoted as design justification: "a sense of shelter," "the invitation of openness," or the "fascination of complexity." Too many statements of this kind are generated by "insight," over a drawing board, rather than identified through research. The values that count are those that are derived directly from the participants in any planned setting. This criterion may seem altogether too vague to be useful in design or in judging design, but that is only because the serious, systematic study of human values as a basis for design has been atempted so seldom. Once a designer has gone through the processes described in earlier chapters, he should find himself with very specific guidelines.

A second human criterion for design would provide each participant with a measure of freedom to adapt his personal environment to his own needs. At first glance this may seem to be an anti-architectural concept, a proposal to provide each user with a basic building kit and let him create his own environment, but this is certainly not what it means to me. It is intended to emphasize that the idea of "total design control," in which every aspect of the environment is ordained by the designer, does not mesh with the realities of human behavior. Once the designer accepts this principle, he will find that there are any number of ways in which he can accommodate individual preferences and provide for optional behavior without compromising the need for a fixed structure and fixed utilities. The most difficult part of this idea for most architects to swallow is that it extends to the exterior as well as the interior of a building. The idea that a tenant in an office tower or an apartment has some rights to reorganize his interior spaces to suit his needs is commonplace. The logical extension of this thought, that he should have an equal opportunity to alter the exterior of his space, is completely at variance with conventional practice.

There is no reason that the opportunity to personalize exterior spaces should not be available to those who wish to use it; nor is there any reason that architects should view this as an invasion of their professional prerogatives. While the profession has traditionally held the view that the exterior of a building should express a single unified concept, the alternate view that the exterior of a building should be able to accommodate diversity is just as valid a basis for a design expression and, in my mind, much more human. There is no single way that this should be done. Finding new solutions that accommodate diversity offers the widest scope for ingenuity and personal genius.

The concept of personal expression on the exterior of a building leads to yet a third design criterion, the concept that if a building is supposed to express anything, it should be related to the human activities contained in it. This is admittedly a difficult criterion to use and to evaluate. All manner of semantic difficulties arise in attempting to analyze the term "express human activities." It is no more obscure, however, than the notion of expressing the nature of the structure, the nature of the materials, or any of the other similar ideas that have been espoused by architectural theorists. Any one of these criteria, which are charged with meaning for architects, would seem obscure to someone who was unfamiliar with the terminology of design. The idea that architectural design should express human activities would only seem obscure to someone who has not tried it.

In any event, a building that accommodates the human tendency to personalize their environment on the exterior as well as on the interior will probably have little problem in expressing human diversity. It will express itself. We have no difficulty accepting the fact that the business establishments we pass as we walk along the public sidewalk are generally quite different in appearance. Extending this same right to the individuals and firms on the upper floors shouldn't be hard to accept either. The labels of "faceless, anonymous buildings" and "filing cabinets for human beings" that have been applied to much contemporary architecture could hardly be applied to an architecture that so directly revealed the rich texture of human diversity.

A fourth criterion, which is a matter of process rather than solution, is user participation. I know from personal experience how hard it is to satisfy this criterion. In some cases, where the specific users cannot be identified, it is literally impossible. Participation is so crucial a human concern, however, so vital in both defining an architectural problem and establishing a positive motivation for acceptance, that a concerned designer should make every effort to deal directly with the users or at least a surrogate population.

The last criterion stipulates that a building should communicate what people need to know in a language that they can comprehend. As we have already seen, design does communicate, though the language is so little understood that the message the viewer receives may be exactly the opposite of the message the designer had in mind. Until someone sorts out that complicated problem, the only reliable basis for evaluating communication is to consider the information needs of the users. People normally see what they need to see to serve their immediate purposes. In entering a new room they may see a table as an artifact that will satisfy their curiosity about the status, taste, and resources of the owner. Later, they may see it as a convenient place to put down their parcels, and still later they will see it only as an obstacle to be avoided in crossing the room. Something very similar occurs in our perception of a building in the public places of the city. During our initial contacts, during the exploratory phase when our need for information is the greatest, we are looking for information that will satisfy our personal needs. What is this place? What does it offer me? What kind of people are inside? How do I participate? If the information needs of every category of participant, the habitue's as well as the newcomers, are described in coincident terms, the functional channels of communication will be defined with as much precision as we are currently capable of achieving.

I realize that this list of criteria does not deal at all with some of the phenomena that have been reported by other writers. It ignores the response we experience in viewing some of the great architectural settings of history. Viewing the Parthenon at sunrise may be an emotionally-charged experience, but we bring with us such a freight of anticipation and preconception that it is impossible to distinguish between the effect of the setting and the effect of our preconceptions. It does not deal with the very keen emotions that an architect may experience in response to physical forms. I know that the pleasure I derive in wandering through new cities or in response to the photographs pinned to the cork wall in my office are so colored by a lifetime of conditioning that it is hopeless to attempt to separate my feelings as an architect from my feelings as a human being. Nor does this list attempt to account for the special enjoyment that any one of us may find in the discovery of an appealing new setting. The factors of personal mood and ambience, our complex individual histories and backgrounds, are such a vital part of this experience that it defies systematic analysis.

My list of criteria may not be complete. It is entirely possible that it may be expanded as further research reveals whether or not there is any reliable and definable system of communication in the design elements of our environment. At the moment, the list covers only those considerations affecting design quality that can be related to human factors.

* * * * * * * * *

Discussing design quality without any reference to mass, scale, proportion, emphasis, or contrast is a considerable departure from the conventional approach to this topic. It would appear to ignore completely all the beloved arcana of design, the legacy of centuries of concern with its basic elements. This is not exactly the way I see it. It is my thesis that the fundamental concern of the planning professions must be shifted from a preoccupation with professional values to solving human problems as these are determined by behavior research. Once the problem is resolved in these terms, the architect must still use such devices as contrast and emphasis to explain and clarify, to make sure the user finds the information he needs. If a user needs such practical information as how to get into a building (which can be something of a problem with some current design concepts) the use of emphasis can clarify the entry. If he needs to distinguish between elements, then form can make the distinction. If he needs to grasp information quickly and accurately, contrast and the associated elements of texture, color, and value, will make it much easier. In each of these cases, however, the special knowledge of design is employed to serve a human purpose rather than the aesthetic purposes of design itself.

It would also appear that this view completely ignores traditional concepts of aesthetics, a point I am willing to concede without any regret at all. It is not that I view architecture and planning as being solely concerned with function. It is my belief that the criteria that have been described can and should lead to a new aesthetic and a new basis for aesthetic judgment. Certainly the knowledge that most perceptions of our environment occur during exploratory movement substantially alters the basic framework of design assumptions. If architecture ever registers as an art form, it is to different people at different times and through a very narrow time window, The studies that reveal the wide gulf that separates the designer as message sender and the viewer as message receiver, and the further indication of the kind of simple, self-serving message to which most viewers are receptive, obviously have all manner of implications both in developing and evaluating new forms. If traditional architectural forms have meaning only to architects, those forms that would convey meaning to the public will rather obviously have a different nature.

Finally, the idea that a building can be evaluated, not through pictures, and not in the exploratory mode, but only in reference to the experience of the users of all classes, must ultimately alter the nature of design judgment and the formation of taste. It may seem strange to include here a reference to taste, which is more usually associated with niceties of fine distinction in all forms of art and expression, but taste is a flexible thing. Taste is a convention, a form of social etiquette, and can literally be based on whatever criteria we elect. An aesthete might argue—as some of them have—that any art form that reflected public criteria would, by definition, be vulgar and banal. This viewpoint is reactionary amd merely indicates that the limits of convention have been breached and new territory liberated. It is perfectly possible for taste to reflect a concern for human values as well as the subtleties of form.

Human values constitute a complex basis for judgment, of course. Such a basis does not permit easy evaluation by glancing at a picture or a casual strolling through a setting. It requires both a precise knowledge of the problem and an accurate assessment of the consequences. From a humanistic point of view, however, it is the only kind of judgment that makes any sense at all.

Chapter VII

How The Architect
Got That Way

In the course of describing a process that should lead to the design of environments more effective in human terms, there is one terminology problem that has been consistently sidestepped. Throughout the preceding pages there have been persistent references to the "design professions" and the "planning professions" without any attempt to define what these terms mean. Such a loose usage is defensible as a convenience, but it is also true that in practice there is no clear-cut distinction between these groups. Just as the term "behavioral sciences" has been employed to describe a whole group of disciplines in the field of human behavior, the term "design and planning professions" has been used to cover a group of separate but related disciplines that are involved in shaping the physical environment.

In practice the term "design professions" is rather loosely applied to those groups that are engaged in the design of physical forms where appearance is a conscious and crucial element. This would clearly include architects and landscape architects. There are other allied practitioners—interior designers, graphic designers, furniture designers, industrial designers, and color consultants—who may not meet the legal definition of "licensed professionals," but who do function in much the same area. "Planning professions" is generally understood to mean those groups that are primarily engaged in arranging and relating the elements of the human environment. The term again includes architects and landscape architects and also embraces city and regional planners, traffic and highway engineers, and engineers who are engaged in planning utility and drainage systems. Either one of these lists could easily be expanded, but as they stand they should be sufficient to indicate the meaning of these two terms.

While simple labels may suffice to distinguish between these two general groups, they will hardly serve to distinguish between the individual practitioners themselves. There is considerable lateral mobility across disciplines. Rather than being bound to his designated professional practice, a landscape architect may become an important member of a highway planning team or the director of a city planning department. An architect may follow a similar course, or may even shift out of design entirely to become an administrator of a municipal building enforcement agency or an executive managing a corporate building program. The conventional, stereotyped role of the architect as an individual with special design and planning skills, operating an independent professional office in the service of clients with building needs, is inadequate to cover all the roles and career paths that may be elected by someone with architectural training. The same thing can be said about any of the other disciplines that have been listed.

In writing about such a varied cast of characters I realize I am running a great risk of making inaccurate observations about professions other than my own, but that is certainly not my intent. In the course of modern practice, with its multitudes of consultants, I have had prolonged contact with all the disciplines that have been discussed and half a score of others that haven't been mentioned. I recognize, however, that such contact knowledge hardly qualifies me to speak for all of them. Even my knowledge of my own architectural profession, which has been intimate over a long term, doesn't qualify me to speak for all of my colleagues. Indeed, it it this intimate knowledge that permits me to state flatly that no one can speak with authority for this group of individualists. As a consequence, in discussing the special attitudes and problems of the design and planning professions. I have focused on architects and, to a lesser extent, on planners in terms that reflect the limited amount of study that has been done on these groups. I would strongly suspect, however, that this information would apply to any of the other groups in the same field. They are all professions, or aspire to be professions, and as such they share some similar characteristics.

In a study of the characteristics that are common to the established professions, William J. Good, a sociologist at Columbia University, found them to be of such a nature that each group could accurately be viewed as a community, or, as other writers have suggested, a sub-society. The members of a profession are bound by a sense of identity, they speak a common language that is only partly understood by outsiders, they share values in common, and the community exercises some degree of power over its members. While they do not produce their next generation biologically, through one kind of selection or another they do exercise a substantial degree of control on who the next generation will be. Through community action they are able to establish standards of education and performance for their members and to influence legislation relating to their profession. It is largely as a result of this unity that the professions enjoy larger incomes than the average occupation in the containing society as well as special status and privileges. Professor Goode points out some of the benefits of unity.

> *The client does not usually choose his professional by a measurable criterion of competence, and after the work is done, the client is not usually competent to judge if it was properly done. ..... indeed, the professional group requires its control over its own members precisely because its judgments do not coincide generally with those of clients. As a consequence, its members need the protection of the professional community and will submit to its demands.*

The allegation that is implicit in this statement, that professions establish their standards and monitor their memberships in order to protect privileged positions rather than to further the public interest, will no doubt raise the hackles of every doctor, lawyer, architect, or engineer reading these pages. Yet there are distinct indications that while the structure of the professional communities may offer some benefits to the public, that is certainly not their sole intent. One of the things a profession might do that would greatly benefit the public would be to rank its members in accordance with some objective measure of competence. Such an evaluation, which would only reflect the great range of skills and abilities that exists in any group, is impossible for a layman to make. Regardless of public benefits, it is inconceivable that a professional group would undertake such a ranking simply because it would be guaranteed to destroy the community that is the basis for professional strength. The professions may enforce their own admission standards and may, within the family, recognize individual differences, but, so far as the public is concerned they are loath to make official distinctions about professional competence. In this sense there is some truth in George Bernard Shaw's acid observations that every profession is a conspiracy against the laity.

Dr. Goode's thesis that the professions can be viewed as a community with special values and role relationships is supported by several studies of the planning professions. John Lansing and Robert Marans of the Survey Research Center at the University of Michigan undertook to measure the correlation between a planner's evaluation of neighborhood quality and the evaluation of the people who actually lived in the neighborhood. They sampled ninety-nine clusters of four dwellings each, randomly selected within the Detroit region, which were then evaluated by a planner on the basis of openness, pleasantness, and interest. These are terms that are well-known to planners and embrace a set of observable characteristics that are highly valued by them—the sense of enclosure or spatial definition, variety in style and topography, architectural interest, and the level of maintenance. Residents in each of the dwellings were then interviewed to determine how they liked their neighborhood and whether they considered it attractive or not.

The disturbing results of the Lansing-Maran study indicate that the planner's value system is sharply different from the resident's. On only one point did they concur; both consider the general level of maintenance in the area as an important index. Beyond that they diverged so decidedly that eighty-eight percent of the people who lived in areas that the planner judged unpleasant liked their neighborhoods at least moderately well. This degree of divergence is so extreme that it implies a complete reversal in terminology between the two groups. The authors, too, were struck by this disparity:

> *At the present we find it difficult to understand why there should be such differences in evaluation. Planners for the most part are well indoctrinated with the belief that enclosed areas convey a sense of well-being and satisfaction to neighborhood residents. Yet a sense of spatial enclosure in fact may not be what people want or deem desirable in residential neighborhoods.*

Studies like the above, and like the Hershberger report on the difference in evaluation between architects and non-architect, underscore the wide chasm separating the value systems of these professions and the value systems of the clients they serve. These findings are difficult to credit unless one is aware of the role-relationships in these fields. In their study *Roles of the Planner in Urban Development,* Robert Daland and John Parker defined four primary roles typically played by a city planner: leader of his departmental organization, professional planner, instigator of political innovation, and promoter of citizen education in planning. In each of these roles he is responsive to different reference groups. In his role of political innovator his reference group is the community leadership structure, but in his role as professional planner his reference group consists of his professional peers. Different value systems color his own appraisal of his performance. If his fellow planners place a high value on comprehensive long-range planning as the mark of an able planner, while his community leadership group demands his attention to more immediate short-range problems, there is a direct role conflict. To succeed at one may mean to fail at the other, and to fail in the eyes of your professional peers is to fail as a professional with the attendant loss, not only of prestige, but of such pragmatic benefits as job preferences and job mobility.

Very similar role problems beset the architect. While I am not about to categorize the members of my own profession—individuals who hold such widely divergent views on purpose, design, social role and life style—there are a few characteristics, at least, that are common to all those who are engaged in private professional practice. They all view planning as superior to not planning, and they all have one crucial reference group—the client. I might have added as a third characteristic that they all hold the view that design quality is an important factor in human life, but at the moment, I'm not certain that this would apply universally.

The view that planning is superior to not planning is one that hardly needs to be defended. It is simply an alternate statement of the proposal that we are more likely to solve our problems by rational analysis than by haphazard individual action. By planning a parking lot with properly sized stalls and clearly defined circulation spaces we can accommodate more cars with greater safety and convenience than if we invited the public to park at random. By planning our public places in a coordinated manner we can ensure the greatest use by the greatest number during the longest span of time. I will concede, of course, that while planning is potentially superior to not planning, it doesn't always work out that way. Much depends on the intent of the planner. The architect who meticulously plans his building so that all the structural components are rational or all the surface elements are modular may at the same time produce interior spaces that are so poorly suited to the needs of the users that they might reasonably have some distinct reservations about the virtues of his kind of planning. On the whole, however, the profession can point to some very substantial accomplishments as a result of rational planning.

The relationship that exists between an architect and his client is unique in many ways; it is also responsible in part for some of the observed deficiencies in our human habitats. With some exceedingly rare exceptions, an architect cannot practice his profession without a client. He can dream the boldest dreams and hone his skills to perfection, but without someone who will furnish the funds to build his concepts he is limited to paper creations—an artist perhaps, or a theorist, but not an architect. This limitation is shared to some extent by other professions; a doctor can't practice clinical medicine without a patient. There is one aspect of the architect-client relationship, however, that is radically different. In many instances the architect's client, who is indispensable to his professional existence, is not the same as the user of the architect's product. When an architect is working directly with an individual in the design of some facility that the individual himself will use, client and user are one, and the situation is very similar to the relationship between a doctor and his patient. When the architect's client is a land developer, a school board, a corporate executive, or even the building committee of a church, the situation is completely altered. These individuals may never even see the finished product or may have only the most superficial contact with it. Yet they are the architect's most influential contacts, his primary reference group. The actual users, who are most directly affected by the results, may have no influence at all.

I don't want to imply that as a result of this separation of client and user that the architect is indifferent to the users. That is not the case at all. Nevertheless, there are certain dynamics involved in role relationships that are inescapable. In bluntest terms, a reference group is one that has the power to hurt you, even if it is only by withholding its approval. In the actualities of practice, the client is near and in sharp focus while the users are distant and indistinct. Unless some means is employed to raise the users to the level of a distinct reference group, they will inevitably be shortchanged.

The whole question of the part played by the client in determining the nature of our built environment is so important in understanding the process that we must return to it later, but first it is necessary to consider how the value systems that characterize the design and planning professions were inculcated.

The Mind-Bending Process

In tracing the process that forms the special value systems of members of the profession it is necessary to start with the schools of architecture. Robert Hershberger's study at the University of Pennsylvania that I cited earlier indicates that the schools are remarkably effective agencies for conditioning the professional mind. In evaluating the reaction of architectural and non-architectural students to the same set of graphic displays, Hershberger also studied a group of pre-architectural students. The pre-architects had the same responses to the displays as the non-architects, suggesting that the responses to forms that they brought with them when they entered architectural school were those of the lay public rather than an innate sensitivity to the design values of the profession. Since pre–architects were separated from architectural students by only a matter of a few years of special training, it is apparent that those few years produced a remarkable alteration in their response to visual stimuli. The measure of this alteration is that approximately thirty percent of the time when the architects judged a building to be good, the lay group judged it to be bad.

While the architectural schools in this country are all directed to the same general end, there is a considerable difference between them both in emphasis and rigor. Within the sub-sub-society of architectural education, some schools are viewed as being rather prosaically "practical," others are considered to be strongly "design" oriented, and still others are rated as highly theoretical. These descriptions vary widely, of course, depending on the source; each architect-educator generally views his own school as being rather precisely focused on producing graduates who are capable of dealing most effectively with the future needs of our society. Within the same circles there is a widely quoted aphorism that "A" students teach and "B" students wind up working for "C" students. While this statement is by no means correct, it does indicate some awareness that the evaluation which schools place on architectural performance may be somewhat different from the evaluation made by society as a whole.

Whatever differences in emphasis or variations in technical and general education courses there may be between schools, the core of architectural education is always found in the design lab. This is the center of the conditioning process. While there are a number of alternative ways to conduct a design lab—using team teaching, and group projects, rather than the more traditional one to one relationship of master and apprentice—the heart of the process lies in solving some simulated design problem. The description of the problem may originate with the instructor, with a committee of instructors, or it may be generated by the students themselves. The students are largely responsible for gathering the information they need to solve the problem, though the instructor may schedule visitors or lecturers from other departments or from outside the school to serve as information sources. As the solution progresses it is reviewed and criticized by the instructor and it is finally presented as a completed exercise in some graphic or model form.

During his extensive study of the sub-culture of architecture, Robert Gutman went through a problem sequence in a design lab and expressed his amazement at some aspects of the process. Working with the same information that was available to the students, he found that he lacked an adequate basis for making the design decisions that were required. This inadequacy, which troubled him as a sociologist, didn't seem to inhibit the architectural students in the slightest. Lacking information, they simply made whatever assumptions were necessary and proceeded with the assignment. Since the evaluation of their work would come from the only relevant reference groups available to them, their instructors and their peers, and since these groups shared the same misinformation or lack of information, the merits of their assumptions could hardly be subjected to a rigorous test.

Dr. Gutman is not the only person who has commented on this aspect of the design lab approach. As it now stands, it is accepted as a proven and reliable way, though not necessarily the most efficient way, to develop the ability to integrate a large number of variables—some of them highly abstract—into a comprehensive solution. This integrating, synthesizing ability is no mean talent; it is, undoubtedly, the most valuable skill the architect has to offer society. It is preposterous, however, to assume that this ability has a social value of its own regardless of the data that is synthesized. The insidious and unintended side effect of the design lab approach is that since the student lacks much of the information he needs, operates free from real-life constraints, and most important of all, has no specific reference group of users, he undergoes a continuous and intensive training in making assumptions about people. Since the evaluation of his assumptions comes largely from his instructors, the only feedback available to him comes from within the architectural system; it is a closed circuit.

The practice of acting on the basis of easy assumptions about other people is not confined to the architectural profession, but wherever it occurs it is highly hazardous and error-prone. Very few readers would care to make crucial decisions about the taste, preferences, values or lifestyles even of close friends, let alone utter strangers. We are much too guarded in our relationships with one another to make this a safe procedure. Yet architects make such decisions all the time. Their facile assumptions—based on introspection—stem from a training in doing just that—making assumptions.

This problem is one that many educators in the design field are concerned about. It is the basis for a growing emphasis on the behavioral sciences within the architectural curriculum, joint appointments to architecture and social science faculties, and joint degrees such as architecture-psychology. These are very encouraging moves that should alter strikingly the point of view of the next generation of architects. There is some reason for feeling, however, that unless the nature of the design lab is altered significantly in the process, the results will be somewhat less than they might be. Robert Gutman tells a tale that illustrates the problem.

In meeting with a group of students who had been given an assignment to design a public housing project in an urban ghetto, the question of how the open space on the site should be allocated was laid before him as a visiting expert. This space could be divided into fenced gardens attached to each dwelling unit so that each family could have a private yard, it could be kept in a large central park where it would be available to all the residents jointly, or it could be arranged so that it would be equally attractive and useful to the surrounding community as well as the project tenants. The question that was posed, and it is an intriguing one, is "Which solution should be adopted?" Robert Gutman, I am certain, could provide the students with a rich background of information that would bear on this point, but I am equally sure that no one could provide the crisp, definitive answer they may have expected. The question is obviously loaded with value implications. It is easy to see that these three different approaches could produce decidedly different results from the standpoint of the individual families involved, their relationships with their neighbors, and the attitude of the surrounding community to the project.

It is encouraging to know that this group of students was aware of the nature of the problem and recognized that the alternatives should be weighed in social terms rather than as purely design decisions. It is nevertheless disappointing that they attempted to arrive at a decision in the design lab through a process of intellectual evaluation. The idea that the ghetto residents might have something useful to contribute, that the question might be seen in an altogether different light as the result of field work with either a real or surrogate population, was not even raised. So long as the idea persists that introspection, however well intended, is an adequate alternative to direct exposure to a sample of users, the design studio will continue to train students in making assumptions.

Once the student leaves school and enters one of the several careers open to him he encounters a new set of pressures. Clients, building codes, construction budgets, and time schedules—factors that are exceedingly difficult to simulate in the design studio—become very real and pressing concerns. There is one problem, however, that remains the same in school or in practice. That is the fact that while the architect's work is directed to the construction of full-size, three-dimensional, physical forms, his concepts are evolved and are ultimately communicated on paper. This introduces a semantic problem that must be common to any field in which the end product is so large that it can only be efficiently studied and defined at some smaller-than-life scale. Obviously, the term "semantic" is not used here in its familiar sense but in the larger meaning of the relationship between symbols and the things the symbols refer to. The design professions use an extensive symbol system, expressed on paper, to represent the planning and design elements with which they deal. As the semanticists have repeatedly pointed out, it is very easy to fall into the trap of assigning significance to the symbol itself and to lose sight entirely of the reality for which the symbol stands. This is a constant problem for the architect. Planning can easily degenerate into an exercise in manipulating symbols into attractive or dynamic patterns without much direct concern with the fact that these patterns, when ultimately translated into walls or walkways, will directly affect many people. The results of this kind of planning can sometimes be seen in plan representations that show a precise relationship between elements

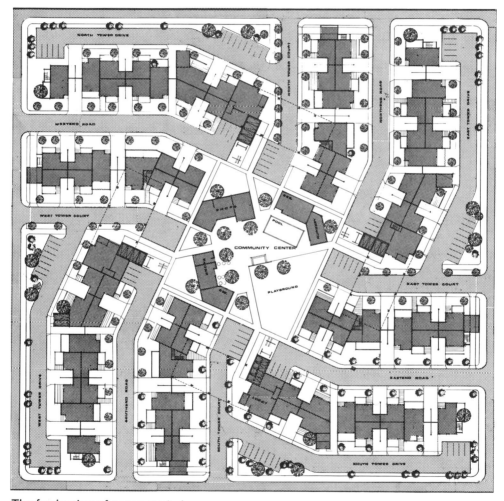

The fascination of paper symbols:

a.   A plan of a hypothetical housing development that shows a pinwheel arrangement of streets, the result of rotating the community center square within the larger square of the total project. On paper this arrangement is interesting but the scale of the project is such that in reality it would be impossible to ever see it.

that in reality may be so far apart, and so obscured by intervening structures, that they are impossible to perceive in any mode of human movement. On paper, however, they may have a powerful attraction.

The source of much of this difficulty is that the architect experiences a very direct feedback from the symbols he organizes on paper, whereas his human constituents can provide no feedback at all during this creative period. Once you have been conditioned to the importance of form and pattern as a measure of architectural merit, it is very difficult to abandon these criteria, particularly when they are performing right before your nose on the drawing board. What the designer is seeking at this point is an ideal combination that will precisely accommodate the needs of the users, satisfy the practical criteria, and meet his own personal standards of what constitutes outstanding architecture. What he is searching for is what anthropologist Edward E. Hall calls "congruence," that happy state of affairs when everything works perfectly for everyone, and a new standard of performance is set. Just how the designer knows when this euphoric event occurs is a mystery, but there is no question in my mind that he does know. Milt Zolotow, a graphic designer, describes it as the "Ah-hah" reaction. When at some point all the elements come together in a perfect union you exclaim, "Ah-hah, that's it." So congruence is born.

Before leaving the Zolotow Reaction I should emphasize that *true* congruence can only occur when all the design criteria are met in an actual setting of human use. For the designer to become enraptured with the arrangement of the semantic symbols on his drawing board and to assume that they are an end rather than only a means is not enough.

A conditioning effect that an architect experiences throughout his career results from the subtle pressures of his own professional reference group. While there are many recognized specialties within the profession and individual architects have made substantial contributions in such specialized areas as technology, building systems, and construction management, there is no doubt that design ability, as reflected in that elusive element of "design quality," plays a paramount part in the internal ranking and recognition of architects within the profession. To some extent this is inevitable. Appearance is the most immediately apparent aspect of any design or planning project and it is the most easily judged and reported. Evaluating a building in use as a setting for human behavior is a long and involved process that the profession has never seriously attempted. Evaluating a building on the basis of pictures and plans, on the other hand, is quick and easy and meshes perfectly with the architect's training and background. From school on, the architect expresses himself through the medium of pictures and plans, develops great skill in interpreting semantic symbols in this form, and is accustomed to evaluate his own efforts as defined in the pictures and plans he continually makes and revises during the design process.

As a consequence, the various design awards programs sponsored by the architectural profession are based on conveniently assembled graphic displays. The results are publicized in the professional press and communicate very clearly that the profession tends to evaluate the artifact itself rather than its effect on the human users. This same convenience factor influences the actions of both the professional journals and home-town newspapers; a photograph of a building is quickly communicated and easily printed while a significant analysis of its effect on people is time-consuming and difficult to prepare. The press introduces another element to architectural conditioning. It does not win readers by continually repeating the same message. Their success is measured by the amount of new information they offer, with a resulting bias toward the new and unique. As a consequence the journals in the architectural field tend to stress innovation and to stress it in the easily illustrated terms of graphic form.

b.   A view down one of the project streets. Obviously, the pinwheel arrangement is invisible from any normal viewpoint. This is a Deasy and Bolling project and illustrates how easy it is to be influenced by semantic symbols on paper. We worked a long time to get that pinwheel perfected.

The organized architectural profession exerts a conditioning effect that is widely felt. It tends to stress the building as a physical artifact rather than as a human setting, to stress the visible structural form of the building to the exclusion of such important factors as mechanical systems and land use, to view architecture as an end in itself rather than as part of a complex system, and to place a high value on innovative forms.

While I believe that the conditioning effects of the design studio, the semantic systems of architecture, and the posture of the organized profession are the germinal sources of the value systems and the characteristic mind-sets that are encountered within the profession (and make it possible to understand why it has not been more effective in dealing with architecture as a behavior setting), I would hate to create the impression that all my colleagues march in lock step. Far from it. Architects are a group of individuals with distinct personal views on a wide range of subjects. They may exhibit common tendencies but never unanimity.

There is one other factor affecting the work of the architect that deserves mention. Faced with the necessity of making decisions about topics on which he has very imperfect information, attempting to serve the demands of a set of conflicting roles as well as his own standards, his lot is not an easy one. As a result of his own experience, Robert Gutman enunciated this problem in a statement for which the profession might well be grateful.

> *Perhaps it is unnecessary to point this out, but let me say that my experience this past year has been exhilarating but also humbling in many respects. For the first time, I developed an appreciation of how difficult it is to be a good architect. Also, I became aware how often the architect is required to make decisions based on information which, to a person with sociological training, seem impossibly fragmentary and incomplete.*

The Party of the Second Part

When an artist creates an easel painting he usually is not constrained to work in reference to a specific user or even to assume that it will end up in a specific setting. He is free to follow his own personal star wherever it leads him. If no one chooses to acquire it, if no one is interested in receiving the message or the emotion he is attempting to communicate, he may be disappointed and frustrated but this does not alter his essentially inner-directed mode of operation. For the environmental designer the picture is radically different. Considering the scale at which he works and the economics of construction, he is generally constrained to work in association with a client. There are a few exceptions to this rule, architects who design, build, and then sell their finished product, but they are rather rare. Robert Sommer has suggested, perhaps sardonically, that architects should consider working as the painters do, originating designs and then selling them to anyone who wants them. This approach is not completely unknown in the field, as evidenced by the existence of plan books, catalogues of pre-fabricated buildings, and the hypothetical projects which architects design from time to time as a means of delineating a personal insight into some problem of the day.

On the whole, however, the architect, landscape architect, or planner is not free to conceive and execute his solutions. There must always be some individual or agency that perceives a need and commands enough resources to satisfy that need. Such an agency is virtually mandatory if the environmental designer is to function at all, yet the nature of their relationship produces some peculiar problems. If an architect is engaged to prepare the plans and supervise the construction of a new school, just exactly who is the client? Is it the pupils, the teachers, the administrators who direct the building program, the elected members of the school board, the surrounding community, or some vague abstraction such as the institution itself? The obvious answer, and the correct one, is that all these groups have a valid stake in the outcome of the project and all are, in a sense, the architect's clients. As a practical matter and as a legal matter, the situation is quite different. The formal client the architect is legally bound is the agency that contracted for his services, that approves or disapproves his work, and that has the sole authority to commit the funds that are necessary for construction. As a consequence, formal clients are a potent group, well-armed with teeth. Regardless of the compassion an architect may feel for pupils and teachers, or the sensitivity he may have about the concerns of the neighborhood, these groups are somewhat remote and often voiceless. If he perceives that their interests are not being well served, his only recourse to alter the situation is to marshal his evidence and exercise his persuasive powers. He has no power to act unilaterally but only with the approval of the formal client.

In spite of the authority that lies in the hands of formal clients it has not been my experience that they tend to be callous or indifferent to the needs of the users if these can be precisely defined. They are as fully committed to the success of the project as any architect could ever be, if not more so. To the developer of an apartment complex, the response of potential tenants to the living quarters he offers is a crucial concern. The operator of a restuarant or even a pizza parlor not only wants to attract customers, he also needs a kitchen where his employees will be both effective and content. Even in the operation of giant corporations or vast public agencies where the eventual users of a building may be extremely remote and utterly voiceless, it is almost certainly their institutional policy to accommodate the users though they may have only the haziest ideas about who they are. Any responsibility that clients bear for the deficiencies of planned environments, especially in human terms, stems not from lack of good intentions but from the special pressures of their position and lack of accurate knowledge about the human factors involved.

Clients have their pressure sources too, as we all do—some higher echelon of authority, a group of elected officials, a Board of Directors, their own peer group, or, perhaps, the friendly loan officer at the bank. Because of these pressures, which may not be visible to anyone else, they may arrive at decisions that amaze the architect but which are completely logical from their own point of view. From the architect's perspective it is like the astonishing sight of an iceberg moving majestically against the wind driven by unseen currents far below the surface. Based on the evidence you can see it doesn't make sense, but it happens.

One frequent source of conflict between the goals of the formal client and the needs of the users stems from the question of longevity; individuals may come and go but the organization goes on. A hospital administrator who may be under great pressure to hold down the cost of hospital services might very well feel that being overly concerned about the views of the nurses with regard to the planning of the nurses' stations would be somewhat irrational since, over a period of time, there will be lots of different nurses using these stations. The corporation executive who is aware of the bewildering speed with which change occurs in administrative organizations may see little point in worrying about the present occupants of a set of offices since he can forsee that in a short time both the occupants and the offices may be gone.

These attitudes have a deceptively realistic air about them but, since they tend to submerge the needs of the people who are involved, they are, in fact, highly unrealistic. While designing a classroom around the idiosyncracies of an individual would be an unrealistic luxury that few could afford, designing classrooms arount the needs of teachers is not. The tendency to stress the corporate or institutional point of view as the decisive factor in determining the essentials of a building program or in establishing the standards that are to be applied is extremely deceptive. Corporations and institutions don't have points of view. Only individuals have points of view. As a practical result, the standards that are invoked inevitably reflect the judgment or the personal concerns of some individual or group of individuals acting as the formal client. If these standards were evolved through some systematic study of human needs, such as the process that has been described in earlier pages, they would have some claim to authenticity, but it has been my experience that this is rarely the case.

One factor that may induce a client to follow a course that is unresponsive to user needs is his personal stake in the decisions that have gone before. An administrator of any on-going building program has made a host of decisions concerning facilities, design standards and construction budgets. Under the circumstances he may view any new evidence that would tend to cast doubt on the merit of his earlier judgments as being distinctly threatening since it might raise some doubt in the eyes of his superiors or his peers about his personal competence. Fortunately, there are clients with sufficient self confidence to realize that the changes that result from a better, broader understanding are a measure of growth rather than a mark of failure; without them we would never have been able to pursue our studies of the human factors in design.

There are other instances in which the client may be circumscribed by the difficulties of changing his own system. This is especially obvious in very large organizations where standards and procedures are established through a long chain of command. The possibilities of improvement may be perfectly obvious to the individuals who are directly involved, and they may be completely in accord with the architect's proposals to accommodate the users or the public in general, but they may also feel overwhelmed by the problem of altering standards through a chain of command that may reach to the state capital, to Washington, or to some corporate board room in New York. In such instances, it is highly probable that the users will lose.

Somewhat different, but still closely related, are those cases in which the reluctance to change in the light of perceived needs stems from the possibility that the change may imply some alteration in the organization itself. This is seldom identifiable as such. It is always possible to advance some other reason for resisting the change, and though the circumstances may raise some dark suspicions in the architects's mind, he is never able to give them substance. We have had at least one instance in which this certainly appeared to be the case. The problem was the design of a library for an extensive public library system, a design to be based on a carefully spelled out program of the facilities to be included. One of the stipulations was that the general reading area was to be divided into two sections, two-thirds for the use of adults and one-third for children. Prior to beginning the design we undertook a somewhat limited observation program in the existing units of the system that reflected the same allocation of space. Even this limited study revealed a wealth of information about the activities of library patrons and the problems of the library staff that raised some serious questions about the appropriateness of some of the concepts that were embodied in the official program. One of the most obvious facts was that the use of these libraries was not split between children and adults on a one-third, two-thirds basis; these libraries were overwhelmed by teenage students who, with complete impartiality, filled the adult side and occupied the tiny chairs in the children's side with their knees under their chins. This evidence raised a number of questions bearing on the nature of the building we were supposed to design. Were the teenagers all that studious or was the library functioning primarily as a social center? If the latter were the case we could provide for their social needs much more easily and economically outside the building than inside. It wouldn't be necessary to turn over a reading room staffed with trained librarians for this purpose. The scarcity of adults raised another question; were they forced out by the teenagers or were they content to select their books and return home? If the latter were true, then we could abandon the idea of an adult seating section altogether and provide more specifically for those studious teenagers. These and other questions were not something we had to conjecture about. They could rather easily be resolved by interviewing some selected samples in the community. In view of the obvious importance of this information in designing a library that would really reflect the needs of the community, we asked our client to provide a modest amount of funds to permit us to conduct such a survey. This request was turned down on the basis of lack of funds. We then asked for a reconsideration of the program requirement dividing the space into children and adult sections, since it obviously did not reflect the actual use. This request was turned down, period.

I am not insensitive to the issues that I believe were involved in this library project. The system did not divide just its spaces on a child-adult basis; budget allocations, book purchases, staff assignments, and civil service ratings followed a similar allocation. Careers had been fashioned around this arbitraty split. It is easy to understand how an executive might blanch at the prospect of upsetting that complicated apple cart in order that one branch library might be better adapted to the needs of the users. It provides an example, however, of how the pressure of an immediate, highly visible group, the headquarter's staff, can assume a priority position in planning decisions at the expense of the users. I sometimes wish that the behavioral scientists who delight in belaboring the architectural profession with the charge of being unresponsive to human needs would spend more time in serious study of the complicated pressures and relationships that are the actual determinants of the final product in cases like this.

Any of the factors I have discussed may distort the actions and decisions of clients to some extent, but the most critical factor is that clients are in no better positions than anyone else to make judgments about the nature of environments that will directly affect the lives and actions of users unseen and unknown. It is not a matter of concern, commitment, or empathy. Although he may have the best intentions in the world, without careful research the clients decisions may be no better than guesses, guesses you and I may have to live with. Even those who are closest to the scene or most representative of the community can err in making such decisions. I have seen well documented evidence of this. Perhaps the clearest example arose in our work with the Hooper Avenue School community. In this instance we were working with an advisory group drawn from neighborhood influentials, a serious group that was committed to a project that would serve the needs of the area and at the same time reflect its values and priorities. In reviewing our proposed interview program with them they considered it highly important that our samples be divided into narrower age brackets since they anticipated that we would find distinctly different attitudes as age varied. Their advice reflected their own reading of the community mind but it proved to be in error; the difference between age brackets was statistically insignificant. In addition, we included the Advisory Group in our interview program as a separate sample in order to compare their responses with those that came from the community they represented. Here again we found enough deviation to indicate that we could not automatically assume that the Advisory Group faithfully reflected the views of the whole community. Our advisers showed a positive concern about local control of the school (as opposed to the exercise of authority by a central school board) and considered the use of more black teachers to be highly important, whereas the community-at-large was generally indifferent to these issues and rated them well below other problems that they considered much more important.

I would hate to convey the impression that the service of the Advisory Group was pointless because they did not provide a flawless mirror of the community they represented. Not only did they provide us with much information and with intelligent critiques of our plans, it is extremely doubtful that we could have worked on easy terms with the community without their endorsement. Certainly, if we had lacked the means for conducting a comprehensive study, their judgement would have given us a much more accurate picture of community concerns than any other single source. This experience supports my contention, however, that any individual or group of individuals who purports to speak for large numbers of human beings is bound to err. To expect any advisory group to speak for their community not only invites error, but is also grossly unfair to the advisors, unless they are provided with the research information they need to carry out their assignment.

Clients suffer from one other problem that is hardly unique. They are all human beings, a burden which they share with architects, planners, and the rest of the human race. As a result of their humanity they are influenced not only by the pressures that result from their special role, but also by their personal values, not the least of which are their own tastes in architecture. Many clients would argue, I am sure, that the architect is supposed to serve as an expert arbiter of good design, enabling them to sidestep altogether this whole thorny question. That is only true within limits. Let the architect stray out of certain ill-defined brackets and clients begin to exhibit great concern. They may look to their architect to be systematic and rational in his analysis of their problems and to originate solutions that are uniquely appropriate to their needs, but if the form of the resulting building falls outside of their area of expectation they may have great difficulty in accepting the consequences. We experienced some difficulty of this sort in designing the Cal State campus union that was described earlier. It would have been impossible to ask for a more favorable climate of opinion or a more receptive client group than we had in that undertaking. The use of social-psychological measurements as a basis for design, a process that might have seemed rather puzzling to some clients, was not only perfectly understandable to them, it was enthusiastically accepted. Our Cal State clients were familiar with the terms and techniques of the behavioral sciences and perfectly at ease in dealing with the abstract questions that were used to generate our data. The whole process was a flawless demonstration of systematic data-gathering and rational analysis with the client and architect working in perfect harmony—until the consequences of the data and the analysis began to assume physical form. At that point we began to experience some severe head-winds. The nature of the structure that emerged from this process clearly did not fall within those mysterious, preconscious brackets that defined acceptable architecture within the minds of our clients. At the intellectual level we had no difficulty arriving at agreement but at the emotional level it was another story. One of the minor points that caused a certain amount of consternation was that the wall openings and windows were anything but uniform in size, spacing or alignment. This external irregularity simply reflected the fact that the openings served different purposes in different internal areas and as a consequence evolved in differennt forms. Judging from the response of our clients this aspect of the design fell far outside the bounds of acceptability. The final result is probably more uniform than it should be but less uniform that they would have liked it to be.

In discussing the special relationship that exists between the environmental designer and his formal client, it has been my purpose to indicate that at least part of the responsibility for the deficiencies that are customarily charged against the designer should be borne by the client. In practice, the decisions that define a project are customarily the product of a client-designer team that generates an architectural setting which will be used frequently by someone who is not on the team. The only way to alter this critical imbalance is to shift that vague and indistinct group known as users, or the public, to the level of a visible and well-defined group, preferably one with teeth. While this is somewhat cumbersome but not impossible in small projects, in large scale programs where hundreds or even thousands of people might be affected, there is no way in which they can all participate directly in a conventional decision-making process. It is my own conviction that in such cases the research processes afforded by the behavioral sciences offer the only feasible means of bringing the users into focus. It is true that these methods do not provide them with authority; a bale of multifold computer printout has no teeth. On the other hand the weight of this evidence is hard to ignore. Once having commited themselves to such a research program and thereby having made some emotional and intellectual (as well as capital) commitment to its outcome, neither client nor architect is complete-ly free to dismiss the results.

Some clients and some architects may object to the idea of injecting a new element into an already complicated decision-making process. They may see it as a limitation on their freedom to discharge their responsibilities as they see them. They are correct. It is, of course, a distinct limitation but an absolutely necessary one for anyone who hopes to produce the kind of settings in which human beings can be more effective.

* * * * * * * * *

In view of the important role that planners and architects play in shaping the artificial environments of our towns and cities, the na-ture of the values and preconceptions that they bring to this task are of more than passing importance to the general public. It is the public, after all, that must eventually survive in these confines, and they could legitimately feel a sense of outrage if their interests are subordinated to a set of artificial goals that the planning and design professions have established for themselves. That such a conflict exists is perfectly clear, and the only means of eliminating it is by restructuring the value systems of the professions. There is no possi-bility that this will be easily or quickly accomplished, though the steps to take seem fairly obvious. Some ways of reducing the social distance that separates the members of these professions from their user constituents are required. As a practical matter this means introducing the user as a real, live, breathing element in design and planning training, moving such training out of the studio altogether, both to establish the needs of the users and to determine the response of human beings to different behavior settings. It also means establishing a new standard of recognition within the profession, based on results in human terms rather that on the form of physical artifacts.- Last of all it means that the practicing professions must adopt and use the techniques of the behavioral sciences to bring the needs and desires of all their clients, formal and informal, into sharp focus.

The stakes that are involved in bringing about a change in traditional design techniques are much greater than are at first apparent. It has been argued that architects and planners themselves are relatively powerless in shaping our cities; the real power is exercised by policy makers—political office holders, land-owners, and building developers. In one sense that is perfectly true, since the environmental designer inevitably can act only through the agency of a client. This argument, however, ignores the much more subtle influence exerted by these professions through the establishment of standards.

There are few aspects of life in our cities that are more circumscribed by regulations than building construction and land use. The average home owner may be completly oblivious to the welter of codes and ordinances that limit his options until he begins some simple do-it-yourself home improvement project. At this point he suddenly discovers that the statement "a man's home is his castle" is simply rhetoric. It may be his castle alone to pay taxes on, but the size of the rooms, the location of the toilets, the nature of the electrical wiring system, the materials of construction, and the position of his home on the lot are decidedly not his alone to determine. These and a host of other matters are covered by regulations. At this point the homeowner's reasonable desire to improve his living quarters may run head on into the value system of the planning professions. He can improve his quarters, but only in certain ways.

The needs for regulation to control physical development in a city are completely rational in a number of ways. On some points, such as fire and health hazards, the city-dweller is acutely vulnerable and has no way to protect himself if the municipal authorities don't do the job for him. There are other kinds of regulation, however, that are more a moral judgment on how people should live than a response to some community peril. Zoning and land-use regulations are an example of this type. In essence they decree that unless an individual can afford to own a piece of land of a given minimum size and to set aside a substantial part of that land for required front, side, and rear yard set-backs, he does not meet the criteria for home-ownership in the community. His willingness to get along with less is beside the point. The argument that these standards are necessary for the general welfare becomes hilarious when the same community applies entirely different standards for the use of land in the hills or along the expensive seashore. If there were some magic number that ensured the good life, it might be assumed that people living with lesser standards along the shore or in the hills would exhibit some visible ills as a result, perhaps a decline in manners or morals, a higher incidence of acute claustrophobia, or some other malaise, but that, obviously, is not the case. While it is true that the neighborhoods that have grown up with these standards will hotly defend them, that is not where the standards originated. They originated in the planning professions or their surrogates.

Moral judgments are also found in the regulations that decree the size and nature of spaces that are suitable for human occupancy. Here again the standards are invoked in the name of human decency without any concession to the fact that many humans would like to have a hand in determining for themselves what that term means to them personally. That option, unfortunately, is not available.

Dr. K. C. Rosser, while serving in India on an assignment for the Ford Foundation, found an extreme example of such moral judgments. During his assignment he had the opportunity to study the conflict between the desires of the unhoused inhabitants of the Calcutta streets and the inflexible attitude of the planning authorities on maintaining housing standards. To someone whose normal home is on the public sidewalk, almost any kind of shelter is a tremendous improvement. Running water, electric lights, and flush toilets are nice, to be sure, but if they aren't available a good sound packing case is distinctly better than nothing. The street people showed considerable ingenuity, in fact, in developing shanty communities, and the authorities could have ameliorated their lot considerably by diverting a modest portion of their housing funds to provide community water and community toilets. This they refused to do. Their point of view is summed up in a quotation from the *Journal of the Indian Institute of Town Planning.*

> *The basic standards in housing and planning are arrived at not only from considerations of cost but also from considerations of creating the desirable sociological and physical environment necessary for the healthy growth of the individuals and the community. These standards cannot be lowered, whatever the community, whatever the location, and whatever the economic situation in the country. Deliberate sub-standard housing will defeat the very purpose of housing as it will lead to the creation of future slums. The basic standards must be adhered to at all costs.*

In short, as long as the Indian planners have anything to say about it, the people will live in standard housing or none at all. They will not even be permitted to solve their own problems that the authorities have failed to solve for them. If any reader has been mystified by my use of the term "professional value system," the above quotation should clear it up.

The kind of moral judgments defined by this quotation is not the exclusive property of planners in India, though I have never encountered anything quite so extreme in this country. Nevertheless, anyone in the United States who could not find accommodation in standard housing and tried to provide some form of shelter through his own ingenuity would find himself in the same boat as his colleagues in Calcutta. The difference is not due to official attitudes but rather to the fact that our supply of standard housing more nearly matches our need than India's does.

Setting community standards is not the exclusive province of any one group, but it is one in which the influence of the planning professions is strongly felt. They are the originators of planning theory and they are the professions to which municipalities turn for counsel in setting their standards. The influence they exert in this manner on the lives of everyone who lives in cities is far greater than is readily apparent or generally supposed. As a consequence, the values and goals that these professions establish for themselves is a matter of considerable public moment.

Chapter VIII

Where Do We Go
From Here?

It is almost a formal requirement for books about planning and design to end with some forecast of a glowing future. Just as a symphony or a concerto will follow prescribed patterns, such books inevitably conclude with a dazzling portrait of the wonderful world that would result if the author's opinions were to become a standard of performance. Since it is this author's view that the human species will at all times, under all circumstances, be implacably and irrevocably human, the demands of that format are difficult to satisfy. Given my premise, it is impossible to argue that our society will soon realize a golden era through a calm and reasoned approach to all its planning problems. The evidence, unfortunately, leads to the conclusion that significant change will occur rather slowly and that, as far as people are concerned, the world of tomorrow will be much like the world of today. On the basis of the information discussed in the preceding chapters it would be quixotic to assume that some utopian configuration of the environment will forever eliminate cant, greed, and hypocrisy as factors in human affairs and provide a more abundant and rewarding life for all. All that can reasonably be claimed is that design based in social psychological research will relieve some of the stress that is a concomitant of urban life, eliminate some of the chaos and confusion that mars so many public programs, and increase the options and opportunities that are available to the individual in pursuing his or her affairs.

These may seem to be disappointingly modest achievements to a nation that commands such enormous technical resources and talent and has come to view technology as the ultimate answer to any problem. The comforting assumption that "if the need is pressing enough someone will build something to take care of it," colors our public thoughts and our public actions. It is neatly expressed in the statement that has been repeated, *ad nauseam*, in the past few years: "If we are smart enough to put a man on the moon we are certainly smart enough to _____." The blank can be filled in with any goal you elect, ranging from "eliminating traffic jams" to "keeping the neighborhood dogs out of your flower beds." This point of view, which has led to an endless stream of utopian technical proposals for solving urban problems, ignores the fact that while space travel is an enormously complicated technical problem, the problems of our towns and cities are enormously complicated human problems. An approach that works for one may have nothing to contribute to the other. While behavior-based design offers no hope of the brilliant breakthrough, the dazzling technological triumph that will change the world, it does offer a method for dealing with human problems in the only terms that hold real promise for improvement. The fact that these changes may be modest rather than dramatic doesn't reduce the value of a behavioral science approach. We may reserve our loudest acclaim for the man who ultimately conquers cancer, but that doesn't mean that conquering a more prosaic curse, the common cold, is an inconsequential achievement.

Finding an appropriate label for the design process that systematically incorporates human behavioral research is something of a problem. A precise description, "Design related to human affairs that is based on a collaboration between behavioral scientists and designers" is impossibly verbose. While a variety of terms have been suggested, my own preference is "behavior-based design" which is simple and probably as descriptive as any brief term can be. Whatever the label, the concepts that are embodied in the idea can be stated rather simply:

1. *Most of us spend our lives in artificial environments that influence our actions, our behavior, and our effectiveness. Since they are rarely designed with conscious knowledge of these factors our behavior is influenced unwittingly.*

2. *The nature of these environments is strongly influenced by the values and attitudes of the planning professions and the purposes of their clients. Regardless of their dedication and commitment, these groups have no way of defining human objectives and concerns without some formal system of research.*

3. *The only processes that are available for defining these factors are those developed and used by the human sciences. Behavior research is capable of representing the user in the decision-making process.*

4. *Regardless of any flaws or imperfections in this approach to planning, it at least guarantees that decisions about the environments that influence our*

*behavior will be made consciously and in response to factual data rather than
unconsciously and as a result of personal whim or bias.*

While these statements define the general purpose of behavior-based
design, they tend to make a complicated process appear simple. To
the non-professional reader, it may not be apparent that the entire
subject matter of this book has been concerned with only a small part
of the complicated task of bringing a physical environment into
being. The professional reader, however, will be keenly aware of this
fact. If the whole sequence were divided into information gathering,
goal formulation, concept development, technical definition, and
construction management, then we have considered only the first
three steps. Resolving the technical problems of construction, preparing the voluminous documents that define the project, coordinat-
ing the work of a variety of consultants and administering the
construction phase, activities that consume the majority of the
professional's time and demand his utmost skill, have not even been
mentioned. This omission is not intended as a slight on the technical
and managerial aspects of design. They are also crucial to a successful
project. Nothing can be more perpetually frustrating than a techni-
cally inadequate environment with its leaks, drafts, and failure-prone
mechanical systems. Nor are there any aspects of construction capable of producing more intense personal stress than a failure to complete a project within the scheduled time span and the allocated cost.
Time, cost, and reliability are also normal human concerns, and to
ignore them in order to focus on less obvious behavioral factors
would not be doing anyone a favor.

My reason for concentrating on the first steps in the design sequence is that this is the point at which decisions that are vital to the
human users are made and the point at which the behavioral sciences
can contribute a special knowledge that has been consistently
bypassed by the design fraternity. While it is true that technical
incompentence can seriously compromise an outstanding concept,
no amount of technical brilliance can transform an inadequate concept. Unless a project is initiated with full knowledge of the human
factors involved, it can never hope to realize its full potential as a
setting for human use. Since these factors may be of those types that
are not usually considered to be germane to planning problems, it is
imperative that they be identified in such specific terms that they
cannot be ignored and must be incorporated in the planning goals.

The development of appropriate goals is crucial for any planning
project, large or small, as it is for any aspect of life. The designer of
an office building may exert his utmost efforts to create a sympathetic working environment for the employees, but if their foremost
concerns are getting an adequate lunch in the teeming streets of the
city, finding a place to park their cars, or being afraid for their

personal safety in walking to the bus stop, he may have missed the real problems altogether. Including factors like these may seem an enormous expansion of the designer's responsibility into new areas, but it only illustrates how easy it is to assume the wrong goals. As long as these concerns preoccupy the thoughts and influence the actions of the employees they are germane to the problem and have to be included in the planning goals. The designer's ability to deal successfully with such issues may be limited but it is unlikely that he can deal with them at all unless he is aware that they exist.

One aspect of goal formulation that can be decidely improved by the methods of the behavioral sciences is in identifying all the different points of view that must be taken into account. The failure to do this has long been a notorious weak point in both private and public planning ventures. It stems from a somewhat arbitrary tendency to assume that a given point of view or a given set of users should dictate the "right" form of the solution. As a result, the operator of a store or the director of a park system may assiduously investigate the concerns of his customers or users, as he certainly should, and ignore the concerns of the non-users. An admirable policy, as far as it goes, but it doesn't go nearly far enough. In many instances it is the non-users who hold the key to success. Finding out why they are non-users and what it would take to convert them may have a substantial impact on the nature and success of both private and public facilities.

Examples of this type of tunnel vision seem to occur with depressing regularity in massive public programs that are directed at revitalizing our fading cities. Whether it is freeway construction, rapid transit, or urban renewal, once a virtuous goal has been defined, and a potential group of beneficiaries has been identified, there is a sufficient basis for initiating a program. The fact that there may be a number of other groups affected, with their own unique points of view, is seldom taken into account. In time, these other groups may make themselves heard with sufficient clarity and force to bring about some drastic changes or even to stop the program cold, but for some reason they are rarely accounted for in the initial formulation of the program.

Perhaps the classic example of this narrow approach to goal formulation is the effort of the past twenty years to improve our cities through a program of urban renewal. The premise on which the program was founded was very convincing. The older areas of some of our cities were in an advanced state of disrepair; they were high cost areas in terms of health care and crime prevention but produced very little in the way of municipal revenues. By removing them and replacing them with new structures our cities would at one stroke eliminate some serious problems and strengthen the tax base. Looking back, I can recall clearly that this argument seemed perfectly plausible at the time. Certainly no one with a shred of humanity would want the people in these areas to continue to live in such depressing surroundings or would question the merits of a bright new city, particularly one with a fine tax base. It was not until much later that I recognized the tragic fallacy in this flawless equation; the needs and concerns of the group most vitally concerned, the people who lived there, had been completely ignored.

This realization came when I participated in a guided tour of a proposed renewal area sponsored by our local renewal agency and conducted by a social worker and a member of the planning staff. All of their commentary was objective and perfectly true; the alleys were unpaved, the buildings old, and the streets in poor repair. The impeccable logic of their proposal to convert this ramshackle area into an upper-income housing district had, however, signally failed to convince one interested group—the residents. As our bus moved through the area it was confronted by individuals who dashed into the streets with hastily lettered signs bearing the legends of intense opposition, "We Dont Want Renewal" and "Improve the Area, Leave the People." The planner's logic also failed to convince me. This happened to be an area where I had spent part of my childhood and as our bus moved through the district, I listened with open-mouthed wonder as our guides described in coldly clinical terms the people and places that I remembered with warm affection. All of their commentary was precisely accurate; the antique oil wells were still nodding ponderously in crazy juxtaposition to the old frame houses, the street pattern was bewildering, the vacant lots were still unkempt, and the population was still heterogeneous and low-income. There was no question that clearing and re-planning the area would produce a more attractive physical environment in an orthodox planning sense. There was, also, no question that doing this would produce the maximum amount of trauma, dislocation and stress for the people who lived there.

Here was a crystal-clear example of a major planning project launched without an adequate definition of the goals and concerns of all the people involved. What is even worse, it typified programs that were being conducted across the nation on a similarly shaky premise. I am not arguing that the city, as a social institution, does not have legitmate concerns about traffic movement, health care, fire and police protection, and tax revenues. These are valid concerns that must be reflected in establishing goals. Considering these as the only goals, however, without taking into account the concerns of the people involved, is not only inhuman, it is the worst possible planning. The basic resource of any community is not streets and buildings but human beings

The proposal for a broader participation in goal formulation in public programs is not unrealistic. It is, in fact, exactly what has happened in urban renewal. In those agencies with which I am familiar the user's point of view has become a major concern, and I believe that this reflects a national policy. It represents a very encouraging change of direction. What is regrettable is that it was so long in coming.

The massive program of freeway and throughway construction that has produced so much turmoil in our major cities shares some characteristics with the renewal program. Here the virtuous goal was to save the cities from strangling in a flood of motor vehicles by opening new arteries. Once that premise is accepted, all manner of atrocities can be excused as a means of keeping the patient alive. In the name of principle, the destruction of homes, shops, parks, and neighborhoods is regrettable but necessary. The people who are affected by these losses are, obviously, not consulted in any significant way. Holding public hearings on a freeway route after an inflexible decision has already been made to build the freeway can hardly be considered participation in goal formulation. The real goal formulation in such cases is based on some professional judgement about community needs and directed to serve the interests of one class of beneficiaries, the freeway users.

Probably no single issue has generated more spectacular controversy in the urban scene than the freeway program. This must be credited to the fact that in arriving at their definition of goals the traffic planners have failed consistently to consider the views of all the involved groups. As a result, the freeways are under continuous attack, and in some instances have been stopped cold, with partially completed structures aimed rather pitifully at an empty void. There is some danger that these comments may be construed as an attack on the minds and motives of traffic planners, but that is not my purpose at all. In my view, they have been given an incomplete assignment and inadequate goals with the result that turmoil builds on turmoil. As Albert Mayer, an architect and planner of vast experience and genuine wisdom has observed, "The traffic planners continually hold out the promise that with the next billion dollars they will be able to solve the problems they created with the last billion."

The "stop the freeways" cry that has been raised in so many quarters reflects a growing disenchantment with the costs of the program in human terms. The appetite of the automobile for both running room and parking space is so insatiable that if we continue to uproot homes and people to make way for more cars we may find that the city we set out to save has moved elsewhere. The most frequently mentioned alternative to this demise by erosion is the development of rapid transit systems, a suggestion that may be viewed with some amazement by those who have lived with such systems for any length of time. The virtuous goal in this instance is quite clear: eliminate traffic problems by reducing the reliance on an inefficient one-man, one-car system and substituting an efficient system for moving large numbers of people at high speed. The obvious beneficiaries would be those individuals with homes and offices near transit lines whose needs would be satisfied by one-dimension travel, in and out of the city. I am in no position to evaluate the merits of all rapid transit proposals, but I confess to a certain uneasiness whenever I encounter our old friends, "virtuous goal" and "single-class beneficiary." There are enormous numbers of people who are left out of that equation. Unless their goals are taken into account, salvation through rapid transit has a dubious future. There is too much truth in H. L. Mencken's acid observation, "There is always a well-known solution for every problem—neat, plausible, and wrong."

One of the great virtues of taking non-users as well as users into account in formulating policies for both public and private planning projects is that it is the only means of discerning the relatively slow shift of public attitudes and priorities. It is the ultimate nightmare of every planner that he will march off into the future on a presumably well-marked path only to look over his shoulder and find that no one is following. While that is not an everyday occurrence, it can happen unless some persistent effort is made to maintain contact with the real world. The best current instance of such a dismal outcome can be seen in the college dormitory program. A number of universities and colleges have undertaken extensive on-campus housing programs in the past few decades, phased in accordance with the availability of funds. Some of these programs offer a textbook demonstration of the best current planning procedures. As each set of buildings was completed, careful studies were made of their use so that the next set of buildings could be improved accordingly. At the end of three or four cycles of construction the dormitory prototype presumably approached perfection. It was at about this point that a number of schools found themselves with a surplus of perfect dormitories on their hands. Nothing in the study processes they had employed was capable of unearthing the fact that a growing number of college students were dissatisfied with formal housing programs of any kind and preferred to solve their own housing problems. It is my belief that the kind of social-psychological programming that has been described in the preceding chapters would have identified this trend and possibly defined a new form of housing that would have satisfied both the student and the school.

The problem of social shift is one of which I am particularly aware through painful personal experience. Shortly before begining our work in behavior-based design we completed a major expansion of a maternity home and hospital for unwed mothers. This institution has provided an essential social service for the Los Angeles region for many decades; the expansion was the last phase of a master plan we had worked out years before. As I write this, however, I am a member of a Board Committee that is seeking an alternative use for what is now an almost empty facility. Changing attitudes about birth control, abortion, and single parents have so reduced the need for a sheltered retreat that it is impossible to justify continuing the operation. Looking back on this experience with the remarkable clarity of hind-sight, it is clear that when the expansion started there were abundant signs that a major change in attitudes was begining. In view of our subsequent experience with behavioral programming, I am strongly inclined to believe that if we had employed these techniques as a routine matter the investment in excess facilities would have been avoided and the staff and administration would have been spared at least part of the stress and anguish of the past few years.

This last example points out one aspect of behavioral science research in the planning field that produces some curious dilemmas. In the strict sense, data about motivations and attitudes does not deal directly with buildings or plans at all. It refers to relationships between human beings and the systems and organizations that have been evolved to accommodate these relationships. In the case of the maternity home it was not the building that was at issue but the whole system of responding to a social need, the sheltered care of unwed mothers. A behavioral programming study in that instance would, I believe, have revealed the need for a drastic alteration of the system. With that information it could be readily inferred that the building needs of the system would also need to be drastically altered. Yet it is far outside the normal province of architects and planners to undertake the design of social systems.

An illustration of this same point can be constructed for the more normal world of commerce. In the course of assembling the information for a building program, data may be collected that bear on a wide range of non-building matters: management policies, customer relations, operating procedures, and personnel policies. It would be a rash architect who undertook to advise a corporate executive or chain-store operator on all these questions. Even to suggest it implies a God-like ability to solve all problems for everyone that is too preposterous to be considered seriously. I'm not saying that it hasn't happened, only that it is preposterous. Yet in attempting to provide settings in which human beings can be more effective, ignoring the policies and procedures that have so much to do with effectiveness doesn't help.

This dilemma is not easy to deal with. Developing appropriate physical settings is so complicated that it should satisfy any designer.

He doesn't really need any new responsibilities. It is clear, however, that structures reflect the form of the organizations and institutions they house, just as the shell reflects the conformation of the turtle inside; if the nature of the organization is defective, so is the structure. As a consequence, it is something of a contradiction in terms to talk about "good" buildings as though they could be judged without reference to the appropriateness of the activities they house. The most brilliantly conceived store, office, college dormitory, or home for unwed mothers, is a dubious achievement if no one wants to use it. This seems to point inevitably to some expansion of the application of behavioral programming in defining the operations and role of organizations as well as their buildings needs. While that does not imply that the architect or planner will assume a new role as the arbiter of whole systems of education, health care, merchandising, or housing, he will certainly have some responsibility for raising the questions about such systems that are revealed by the process of behavioral programming.

We are just begining a project that should shed some light on this specific question: how behavioral programming can be used to develop operation programs as well as building designs. The Los Angeles Public Library System, which operates highly successful branch libraries in most parts of the city,has other branches that receive very little use. The source of this problem obviously doesn't lie in the buildings that house these branches but in the services and activities that are supplied. By use of the research techniques of behavioral programming we feel that we can develop programs and services that will permit each branch to adapt to the unique requirements of its own neighborhood. In dealing with such basic questions as the role the branch library should play in each neighborhood, we are obviously moving considerably beyond the architect's traditional assignment; our assignment, in this instance, is more like systems planning than architectural planning. Nevertheless, we are not trying to be librarians. We will simply be providing the expert staff of the Los Angeles Library System with recommendations based on unique information they have never had before. They are obligated to make the system work, and consequently they have the right to decide what the system shall be.

Even without such forays into the realm of systems planning, policy questions can arise with regard to building programs that are difficult to deal with. They arise when different parts of a constituency have different objectives or when they espouse positions that your own experience indicates may be short-sighted. In order to illustrate that problem we'll have to return to California State University, Los Angeles, one more time. In our extensive interviewing of groups both on and off the campus we raised several questions about priorities of interest. All groups were in accord that the new campus union should ultimately reflect the priorities of the students. In line with that directive, when we found some conflict between groups, we adopted the student point of view in establishing our criteria. There was one point, however, where the student point of view seemed questionable. This had to do with the relationship between the school and the surrounding community, and specifically with the issue of community access to the new structure.

In contrast to some urban universities, where Town and Gown develops a strong and symbolic relationship, Cal State had never forged any strong ties with its surrounding community. The view of the school toward the community and the community toward the school might be characterized as friendly indifference, if there is such a thing. The students considered that to be a splendid state of affairs. They had no interest in the surrounding community and could conceive of no reason why their neighbors should be invited to participate in any way in their student-funded project. The administration, on the other hand, thought neighborhood involvement on campus would be a great idea. They were perfectly aware that the support of any public institution and the value of the degrees it offers is largely contingent on community attitudes. Since the neighboring public, had not the haziest idea of the qualities of the school or the excellent programs it offered, as demonstrated by our survey, not only these students but generations of students yet to come might benefit if the neighbors were enticed onto the campus.

At this point some policy decision had to be made. The student view could hardly be overruled. There is no point in surveying opinion if you are going to ignore the opinions you get. Yet it is hard to escape the feeling that, over the long term, their position was very short-sighted. Our ultimate response was to reflect student attitudes faithfully and then to re-examine our criteria to see what alterations would have to be made to make the project adaptable to both student and community use. The architectural result is an arrangement that will permit the students to change their minds if they ever wish to and invite community participation without limiting their own use of the structure. In other words, the solution reflects present attitudes and behavior, but it also recognizes that these attitudes may change and a new behavior pattern emerge.

This example raises a very important question that will undoubtedly become more acute as behavioral research becomes a normal adjunct of planning not only buildings, but whole systems: who should make such design decisions? The behavioral scientist, with a tradition of detachment and objectivity, may balk at the obvious value judgements that are required. The formal client, faced with a confusing array of data and subtle distinctions of values and priorities, tends to rely on professional consultants to chart a course of action. Yet the architect or planner who undertakes to supply these answers may find himself dealing with issues that are far outside his normal area of interest and training. In practice, we have undertaken to resolve as many of these questions as possible ourselves and submit them to our formal clients as recommendations. While they have rarely been challenged, I am not inclined to dispute the possibility that there may be a better way for fixing policy.

* * * * * * * * *

Before bringing this discussion of behavior based design to a conclusion there are a few more peripheral points that should be mentioned. One, not unnaturally, is the question of cost. Carrying out the systematic studies that are necessary in developing a behavioral program cost money, and in large undertakings involving many people it may cost quite a bit of money. While our Cal State study involved a direct cash expenditure of only $15,000, it is easy to see that the large scale studies that might be necessary in order to deal with community-wide projects such as transportation systems would inevitably cost many times as much. While such sums are a minor fraction of the cost of carrying out the actual construction program, they are not inconsequential in relation to the compensation a professional planner or architect receives for his services. As a consequence, the cost of behavioral programming will inevitably represent an added charge that must be borne by someone. The projects discussed in the preceding chapters that reflect our own experience were funded from a variety of sources. In each of these instances we were commissioned to provide the normal range of architectural or planning services and the special social-psychological studies that were required were funded as a separate activity. In three cases the cost was covered by grants from the Educational Facilities Laboratory, an affliiate of the Ford Foundation, or the Council on Library Resources. In other cases private clients were sufficiently persuaded of the value of this kind of information to pay the cost themselves. The last category, largely limited to observational studies, were those we funded ourselves as an internal research program.

There is no doubt that the use of behavioral programming will be severely limited until it becomes generally accepted as a normal part of planning services. In view of the high priorities accorded economic criteria relative to human criteria, this may be a long time coming. Oddly enough, it may come more quickly in commercial fields, where it holds the prospect of commercial benefit, than in those areas of social concern where it might produce the maximum benefit to the community. Nevertheless, there is no one class of client that will automatically recognize the benefits that can accrue from an investment in better programming. Spending money on bricks and mortar produces direct, tangible results; spending money on information about human behavior produces nothing tangible at all, even though it is crucial in arriving at the most effective arrangement of the bricks and mortar. As a consequence, the concept that artificial environments can only be assessed accurately when they are considered as behavior settings is hard for many clients to grasp. There are so many positive, pragmatic benefits produced by this approach, however, that there is reason to believe that it will ultimately be accepted as a planning norm.

There may be some fear that when the public and the staff are involved in the planning process that they will generate expensive demands that will be difficult or impossible to meet. Based on our own experience these fears seem to be groundless. It seems highly probable, in fact, that some of the ideas generated by clients and their architects represent expenditures that the users may view with complete indifference. The Hooper Avenue School community had ample opportunity to express a preference for expensive refinements in their new buildings but they consistently ignored them and concentrated on those elements that had a direct bearing on education.

There is one area of cost, involving the relative ranking of economic and human criteria, that will require some alteration in conventional patterns of thought. In certain classes of planning projects, primarily but not exclusively in public programs, cost criteria of some sort are invariably specified. In view of the demands on the public treasury this is a perfectly rational limitation. In applying these criteria, however, there is always some recognition that they may have to be adjusted in some degree to meet the realities of construction. In constructing a highway, for example, there may be a standard cost per mile allocation that reflects a certain level of construction quality. During the construction of the highway, conditions may be encountered that require this allocation to be revised. Extensive rock outcroppings, marshy terrain, or underground water may so alter the foundation conditions that the unit cost criteria are no longer relevant. To adhere rigidly to the original cost criteria would simply mean that the highway could not be completed. Unfortunately, this same flexibility is rarely available with regard to human criteria. If the highway in our example should pursue its course through an urban area, the added cost of depressing it below the surface to preserve such human values as a reduction in noise, preservation of views, easy contact between neighbors and maintaining the intricate network of contacts that characterize a viable community, would not usually be considered a reasonable expenditure. Human criteria, in other words, do not share the same status as economic criteria. Until they do, the promise of a reduction in environmental stresses is sharply limited.

It can be argued, of course, that if the public were permitted to enter into goal formulation to a large degree many programs of great merit would be completely blocked or so restricted that they could never achieve their purpose. That is a hard question to deal with. I realize that the public can show a distressing tendency to defeat bond issues for essential community purposes such as adequate sewers or decent jails that are somewhat remote from their everyday concerns. I would argue, however, that the public presented via protest rallies, pressure groups, or even the ballot box, is not necessarily the same public that one finds in face to face interviews or through careful observation. I have a strong predisposition, based on some experience, to believe that the public can be surprisingly rational and objective under such circumstances. It would be irresponsible for me to propose that if behavioral programming and goal formulation became a mandatory part of important public programs conflict would be eliminated, harmony ensue, and we would live happily ever after. On the other hand, in view of the conflict that already prevails, it is certainly worth a try. If, on the basis of such a study, the public persisted in rejecting certain programs, it would raise some serious questions in my own mind about the value of the programs or their position on the list of community priorities.

In addition to questions of cost, there is one other factor that may limit the expansion of behavior-based design; there must be someone who is both willing and able to do it. I have consistently made the assumption that this kind of design implied some degree of collaboration between designers and behavioral scientists. In view of my own backround and experience, that is a natural bias. The procedure that ultimately evolves, however, is really contingent on the interest of the behavioral scientists and my colleagues in the planning professions. It may appear to be an arduous role to both of them. I have a keen appreciation of the complexities that are already abundant in the fields of architecture and planning; for these groups to take on an added responsibility may strike them as altogether too much. If that is the case, it may lead to the development of the new discipline predicted by Robert Sommer, a kind of applied behavioral scientist who is equally at ease in directing research activities and in describing practical applications. I could have no real argument with that prospect if it proved to be the only way to get the job done, but I still hold the hope that the planning professions will undertake this collaboration themselves. The planner or designer who undertakes to ameliorate the human condition, to provide settings in which human beings can function at their best, can never get too close to reality.

Introducing a new discipline, a new filter for information, only acts to move him one step further from the people who are his primary concern.

While it is true that the procedures that are an inherent part of behavior-based design are complicated and time consuming, they also offer the designer an effective new tool for improving the human habitat. This new role nessarily implies a new responsibility, but it also provides an opportunity of a scope that is impossible to forecast at this time. No one knows the extent of the stresses we experience through living in buildings and towns that are imperfectly adapted to our use and no one knows what our response to better adapted habitats would be. We can identify bits and pieces of the picture in specific projects—a slight reduction in stress here, an improved opportunity for social contact there—but the total effect is unknown. It may be small, or it may, in the aggregate, be large. In either case, it is an opportunity that should not be bypassed. The designer who is capable of improving human effectiveness in any degree is a very useful member of society.

In all the preceding pages I have attempted to curb my abundant enthusiasm for the concept of behavior-based design and have tried not to imply that it would produce any miracles. There is one observation, however, relating to man-environment matters, that I cannot manage to contain. In all the vast fund of misinformation about human nature that constitutes our conventional wisdom, there is one point that is consistently overlooked: the same people who are capable of being mean, grasping, and thoughtless are equally capable of being kind, generous, and considerate. As a consequence, we frequently think and act on the conviction that they will consistently perform at their worst without considering the possibility that under proper circumstances they may be induced to perform at their best— this in spite of the fact that for every instance of deplorable human behavior another instance of admirable behavior can be cited.

The ecologists who are concerned about the destruction of our physical environment have compiled an overwhelming indictment of the greed, ignorance, and neglect that have so tragically despoiled our planet. Yet, the same human race has created some of the most supurb pastoral settings the world can offer. The beautiful agricultural valleys of Switzerland are not natural settings; they are natural settings that have been adapted to the use of man with care and understanding. The lush farms of Lancaster County in Pennsylvania are an alteration of nature as were the infamous bonanza farms of the Dustbowl, but in Lancaster County the alteration serves the purposes of man only to the extent that it does not destroy nature. As a consequence, while I share the alarm of the ecologists, I at least see some hope in the fact that human beings are not intrinsically destructive as long as they are capable of the attitudes that produced the Swiss valleys or the farms of Lancaster, they are capable of altering their destructive behavior.

In a similar sense it is possible to compile a gloomy catalogue of human behavior in cities: indifferent, self-centered, callous, and grasping. No one need look very far to confirm the fact that these characteristics can, indeed, be found in cities. It is equally clear, however, that they are not an intrinsic part of human nature. Those same citizens who can, on occasion, be so thoroughly obnoxious to each other can also, on occasion, rise to unparalleled heights of generosity and compassion. In view of that fact, I maintain the hope that human behavior in cities is not altogether beyond redemption. Given settings that offer them an opportunity for being at their surprising best and the least cause for being at their unbearable worst, the composite result may be surprising. How much good could be accomplished by a better, more humane environment is impossible to say, but it is a place to start. The possibility of finding out has been ignored too long.

Bibliography

Information about new developments in the field of environment and behavior is hard for the lay reader to acquire. Much of it is in the form of papers presented at conferences (with a very erratic record of publishing in *Procedings*) or privately printed monographs. A library at a university with a school of architecture would be one good source. The Library of the American Institute of Architects, The Octagon, Washington, D.C., would be another.

The following list of titles are those that are referenced in the text and is by no means comprehensive. In addition, the firm of Deasy and Bolling, Architects, has issued reports from time to time on special project studies that seem to have value to our colleagues. Some of these are refferred to in the text.

Social Psychological Considerations in Architectural Planning, 1966. A study of the social and psychological factors involved in planning an office building in downtown Los Angeles.

People in the Streets, People in the Parks, 1967. A study of public behavior in downtown Los Angeles leading to the design of a public plaza.

Actions, Objectives and Concerns, Human Parameters for Architectural Design, 1969. This is the Cal State study for a new Student Center that is mentioned so frequently in the text. The body of the report is a point by point analysis of the information derived from observations and interviews. Appendix A contains the full text of Dr. Thomas Lasswell's report. This study was funded by a grant from the Educational Facilities Laboratory, Inc.

Real Goals Versus Popular Stereotypes in Planning for a Black Community, 1970. This is the Hooper Avenue School Study. The project was to replace the superannuated buildings at a ghetto school. Appendix A contains the full text of Dr. Lasswell's report. This study was funded in part by a grant from the Educational Facilities Laboratory, Inc.

We have sent out hundreds of copies of these reports but they are no longer available. Circulating copies may be obtained from the Library of the American Institute of Architects.

Abrams, Charles. The City is the Frontier, New York, N.Y.: Harper, 1965

Alexander, Christopher. Notes on the Synthesis of Form, Cambridge, Mass.: Harvard University Press, 1964

Alexander, Ishikawa, and Silverstein. A Pattern Language Which Generates Multi–Service Centers, Berkeley: Center for Environmental Structures, 1968

Barker, Robert. Ecological Psychology, Stanford, California: Stanford University Press, 1968

Bechtel, Robert. "Human Movement and Preference" in Proceedings of the 1966 American Institute of Architects Architect–Researcher Conference, Washington University, St. Louis, Missouri

Birrens, James. "The abuse of the Urban Aged" in Psychology Today, March, 1970

Brolin, Brent and Zeisel, John. "Mass Housing, Social Research and Design," Ekistics, 27, January, 1969

Churchill, Winston. "Rebuilding the House of Commons" in Charles Eades Onward to Victory, Boston, Mass: Little, Brown and Co., 1944

Daland, Robert and Parker, John. "Roles of the Planner in Urban Development" in F. Stuart Chapin and Shirley F. Weiss (Eds.) Urban Growth Dynamics, New York, N.Y.: John Wiley and Son, 1962

Deasy, C.M. "People Patterns in the Blueprints" in Human Behavior, August, 1973

Deasy, C.M. "People Watching—With a Purpose" in Journal of the American Institute of Architects, December, 1970

Deasy, C.M. "When Architects Consult People" in Psychology Today, March, 1970

Deasy, C.M. "When a Sociologist Gets Into the Act" in Journal of the American Institute of Architects, January, 1968

Dubos, Rene. So Human an Animal, New York, N.Y.: Scribners, 1968

Festinger, L., Schacter, S., and Back, K. Social Pressures in Informal Groups, New York, N.Y.: Harper and Bros., 1950

Foote, N., et al. Housing Choices and Housing Constraints, New York, N.Y.: McGraw–Hill, 1960

Gans, Herbert J. "The Ayn Rand Syndrome" an interview by R.W. Glasgow, Psychology Today, March, 1970

Goldberg, Theodore W. "The Automobile, an Adolescent Institution Colliding with the Larger Community" in Man–Environment Systems, July, 1969

Goode, William J. "Community Within a Community: The Professions" in American Sociological Review, April, 1957

Gutman, Robert. "Site Planning and Social Behavior" in The Journal of Social Issues, 1966, 22.

Gutman, Robert. "Sociology in Architectural Education" in Proceedings of the 1966 American Institute of Architects Architect–Researcher Conference, Washington University, St. Louis, Missouri

Hall, Edward T. The Hidden Dimension, New York, N.Y.:Doubleday and Co., 1966

Hall, Edward T. The Silent Language, New York, N.Y.: Doubleday and Co., 1959

Hershberger, Robert G. "A Study of Meaning and Architecture" in Proceedings

EDRA I

Jacobs, Jane. Death and Life of Great American Cities, New York, N.Y.: Random House, 1961

Lansing, John B. and Marans, Robert W. "Evaluation of Neighborhood Quality" in Journal of the American Institute of Planners, May,1969

Leighton, Alexander. My Name is Legion, New York, N.Y.: Basic Books, 1959

Lerup, Lars. "Suburban Residential Environments: Analysis, Evaluation, and Selection," Ekistics, 183, February, 1971

Lindheim, Roslyn. "Factors Which Determine Hospital Design" in American Journal of Public Health, 1966, 56, No. 10

Lindheim, Roslyn. "Uncoupling Spatial Systems" a paper delivered at the 1965 American Institute of Architects Architect–Researcher Conference, University of Michigan, Ann Arbor

Lorenz, Konrad. King Solomons Ring, New York, N.Y.: Thomas Y. Crowell, 1952

Mayer, Albert. "Architecture as Total Community" in Architectural Record, July, 1964

McLuhan, Marshall. Understanding Media, New York, N.Y.: McGraw–Hill, 1964

Michelson, William. "Selected Aspects of Environmental Research" in Man–Environment Systems, July, 1970

Milgram, Stanley and Toch, Hans. "Collective Behavior: Crowds and Social Movements" in Gardner Lindzey and Elliott Aronson (Eds.) The Handbook of Social Psychology (Second Edition), Reading, Mass.: Addison–Wesley Publishing Company

Myrick, Richard and Marx, Barbara. An Exploratory Study of the Relationship Between High School Building Design and Student Learning, The George Washington University, Washington, D.C., 1968

Newman, Oscar. Architectural Design for Crime Prevention, Washington: National Institute of Law Enforcement and Criminal Justice. U.S. Government Printing Office, 1971

Newman, Oscar. Defensible Space, New York, N.Y.: The MacMillan Company, 1972

Norberg–Schulz, Christian. Intentions in Architecture, Cambridge, Mass.: The MIT Press, 1965

Payne, Ifan. "Pupillary Response to Architectural Stimuli" in Man–Environment Systems, July, 1969

Perin, Constance. With Man in Mind, Cambridge, Mass.: The MIT Press, 1970

Phillips, Derek L. "Social Participation and Happiness" in American Journal of Sociology 72 March, 1967

Piaget, Jean and Inhelder, Barbel. The Psychology of the child, New York, N.Y.: Basic Books, 1969

Priest, Robert and Sawyer, Jack, "Proximity and Peership: Bases of Balance in Interpersonal Attraction" in American Journal of Sociology 72 May, 1967

Rasmussen, Steen Eiler. Experiencing Architecture, New York, N.Y.: MIT Press and John Wiley and Son, Inc. 1959

Rosser, K.C. "Housing for the Lower Income Groups; The Calcutta Experience", Ekistics, 183, February, 1971

Sauer, Louis and Marshall, David. "An Architectural Survey of How Six Families Use Space in Their Existing Houses" in Proceedings of AR8/EDRA3, University of California at Los Angeles, 1972

Smith, Downer, Lynch and Winter. "Privacy and Interaction Within the Family as Related to Dwelling Space" in Journal of Marriage and the Family, August, 1969

Sommer, Robert. Personal Space—The Behavioral Basis of Design, Englewood Cliffs, N.J.: Prentice–Hall, 1969

Sommer, Robert. The Ecology of Study Areas, The University of California, Davis, 1968

Sumley, F.H. and Calhoon, S.W. "Memory Span for Words Presented Auditorially" in Journal of Applied Psychology, Vol. 18, pp.773-784, 1934

Van der Ryn, Sim and Silverstein, Murry. Dorms at Berkeley—An Environmental Analysis, New York, N.Y.: Educational Facilities Laboratories, 1967

Van der Ryn, Sim. "Ecology of Student Housing" in Proceedings of the 1966 American Institute of Architects Architect–Researcher Conference, Washington University, St. Louis, Missouri

Wehrli, Robert. Open–ended Problem Solving in Design, Department of Psychology, University of Utah, 1968

White, L.E. "The Outdoor Play of Children Living in Flats" in Leo Kuper (Ed.) Living in Towns, London: The Cresset Press, 1953

Young, Donald R. "Behavioral Science Application in the Profession" in Bernard Berelson (Ed.) The Behavioral Sciences Today, New York, N.Y.: Harper and Row, 1964